3000 800011 52136
St. Louis Community College

D1086931

INVENTORY 98

To see, not simply look at:
a phenomenal world perceived by a consciousness which discovers that it is itself a phenomenon. Receptive to an ever-new relationship between the external and internal universes of the body, the senses become a mirror of the energies which surround it, and the witnessing intelligence becomes an instantaneous mirroring of the timeless cosmic spirit...

Optical Color & Simultaneity

Ellen Marx

Translated by Geoffrey O'Brien

VNR

VAN NOSTRAND REINHOLD COMPANY

New York Cincinnati Toronto London Melbourne

English translation by Geoffrey O'Brien
Library of Congress Catalog Card Number 81–4731
ISBN 0–442–23864–9

Originally published as *Couleur optique*
Dessain et Tolra, Paris

Printed in France

Published by Van Nostrand Reinhold Company Inc.
135 West 50th Street
New York, New York 10020

Van Nostrand Reinhold
480 Latrobe Street
Melbourne, Victoria 3000, Australia

Van Nostrand Reinhold Company Limited
Molly Millars Lane
Wokingham, Berkshire RG11 2PY, England

16 15 14 13 12 11 10 9 8 7 6 5 4 3 2 1

Library of Congress Cataloging in Publication Data

Marx, Ellen.
Optical color and simultaneity.

Translation of: Couleur optique
Bibliography: p.
1. Color. I. Title.
QC495.M3313 1981 535.6 81–4731
ISBN 0–442–23864–9 AACR2

Contents

Preface

After finishing *The Contrast of Colors,* I continued to be troubled by further questions, until finally I embarked on a new series of systematic empirical experiments.

Why are there two kinds of complementaries, optical on the one hand and subtractive on the other? Which of the two is the simultaneous complementary? It was successive contrast that gradually provided me with the solution.

What is the difference between additive and optical synthesis? A number of writers, even contemporaries, still confuse these two types of synthesis. Can saturated colors be created optically? Under what conditions does the optical end product contain white or black?

It then became necessary to establish connections between phenomena such as simultaneity and optical color, both of which only exist as subjective visual impressions. In order to discover the exact quantities with which the primary colors (corresponding to the thresholds of vision) as well as the complementaries can be achieved through optical mixing, it was necessary to ascertain the precise percentages by means of two different methods: on the one hand by the rotation of disks; on the other by spending hours pasting up tiny dots with a razor blade, making sure that the dots were exactly equidistant.

Then, for days on end, under constantly changing mental, physical, and environmental conditions, I focused my attention on the most minute variations in the contrast of colors which have no objective existence: the successive images.

Before even beginning this research, it was essential to choose base colors that would make it possible to cover all the possibilities of optical synthesis.

Frans Gerritsen, in his book *Color Presence,*[1] writes: "Blue and yellow, considered as neural input, are opposites; considered as visual colors, they are complementary. Red and green are also opposites, considered as neural input, but are not visually complementary; when added they make yellow rather than white."

The problem here is one of imprecise terminology. Red and green, optically and by addition, create yellow if it is a red of 100% magenta + 100% yellow and a green of 100% cyan + 100% yellow. Red and green are at the same time visually complementary — but it remains necessary to define precisely which red and which green you are talking about.

On page 66 it is shown clearly that the optical complementary of, for example, a red of 100% magenta + 30% yellow has as its optical complementary a green of 100% cyan + 50% yellow.

In *Solfeggio of Color,* Edouard Fer writes: "Conclusive proof can be obtained by the use of a Maxwell disk, where the rotation of a yellow paired with a blue never produces green or anything approaching it."[2] Tangible proof of the contrary will be found on pages 35 through 40 of his book. Cyan-blue when combined optically with a yellow incontestably produces a green rather than a gray. The green will be purer to the extent that the primary cyan-blue is devoid of any trace of red-magenta. The blue optically complementary to yellow (see pages 111 and 112) has a red-magenta component of 50% to 70%, depending on the characteristics of the primary inks.

These are only two examples of the imprecisions in the specialist literature which spurred my desire to clarify the mechanisms of color perception through concrete demonstration.

Finally, you can provide your own verifications with the help of the graduated circle on page 142 and the detachable color disks on pages 144, 145, 147, and 149. By using these you will be able to make your own exciting discoveries.

The Three Systems of Synthesis

When two rays of light are superimposed, the resulting color will be brighter than the brighter of its components. For example, when a blue-violet is added to a red-orange, the result will be a magenta which is brighter than the red-orange. (This synthesis is called *additive* because there is an augmentation of energy.) All the luminosities of the spectrum combined attain the highest level of energy — white. Inversely, when it passes through a prism the white light breaks down into all the luminosities of the spectrum, with the exception of the reds between red-orange and magenta and the red-violets between magenta and blue-violet. (Compare with the two scales on page 37, using the transparent magenta sheets.) For a long time this fact had serious repercussions, not only on the manufacture of dyes but on the making of theories as well. It is only recently that magenta has made its appearance on the art supplies market.

Three radiations with carefully selected wavelengths (i.e. blue-violet, green, and red-orange) on the one hand, or a single pair of complementaries on the other, are sufficient for obtaining a neutral white.

We are dealing here with the subtractive process, which is involved when substances interact. On a white background, every coat of transparent ink superimposed on another diminishes the total energy, culminating in black, which absorbs all the chromatic rays.

In printing, only flat tones are pure subtractive systems. When different screens are superimposed, the points overlap at certain places (subtractive), and at others they influence each other by their proximity (optical). Black can be obtained not only from the three subtractive primaries (yellow, cyan, and magenta) combined, but also from all the complementary pairs.

In color photography, the light rays are absorbed by three layers of filters in accordance with the same subtractive laws. For a transparency, for example, the basic white is not paper but the white light necessary for projection. Depending on the degree of transparence, each of the three filters modulates its complementary: thus, the yellow filter is responsible for the blue-violet, the cyan filter for the red-orange, and the magenta filter for the green.

Optical color is nothing but an impression created in the retina as a result of two or more colored elements. The threshold of discrimination can be exceeded in two cases:

1. When narrow lines are extremely close to each other or when points of color are of such small dimensions that they irradiate onto each other and thus lose their individuality in a uniform colored surface. The new color thus created is an intermediate impression of the totality of the values present. The interdependent relationship between size and distance from the observer must not be forgotten; a surface that at close hand is made up of distinct elements can become a homogeneous surface when seen from a distance. In printing, screens whose point size is less than 1/6mm. give the impression of being flat tones.

2. When a disk turns at a sufficiently high rotation speed (a frequency of about 3,000 rpm) to cause the sections of color (independent when the wheel is not in motion) to fuse into a single impression. The optical combination will produce, in terms of brightness, hue, and saturation, an average of the separate elements.

Whether the fusion is caused by a proximity of minute structures in space or by a proximity of oscillations in time, the result will be the same. It is a process in which there is neither loss nor gain of energy.

Many writers have mistakenly tried to identify optical mixing with additive synthesis. Newton was probably the first to attempt to reconstitute white by rotation of the disk on which he had arranged his seven colors, seven being a magic number suggested more by the superstitions of his time than by any physical reality.

That the product will always be gray rather than white is not due to impurities in the coloring material. It is rather that the mixture of electromagnetic impulses on the organic level, whether coming from illuminated surfaces or from sources of light, cannot be transposed to superimpositions of lights on a screen external to the eye.

Since Newton, a plethora of physicists, physiologists, and painters have made systematic observations with rotating disks, notably the often-cited experiments of Maxwell.

The three diagrams opposite permit a comparison which will enable us to understand the correlations between fields as diverse as optics, video technique (Table 1), chemical manufacturing, especially the development of dyes (Table 2), and the psychophysiology of sight (Table 3). Optical synthesis can be placed exactly midway between the additive system (energy) and the subtractive system (matter), since the basis of sight (eye — conduction of neural impulses — brain) integrates the totality of the impulses into a single impression, representing an average value of brightness, hue, and saturation.

1. ADDITIVE SYNTHESIS: ENERGY

short blue-violet	+	medium green	+	long red-orange	=	white

(wl = wavelengths)

long wl red-orange	+	medium wl green	=	yellow
short wl blue-violet	+	long wl red-orange	=	magenta
medium wl green	+	short wl blue-violet	=	cyan

2. SUBTRACTIVE SYNTHESIS: SUBSTANCE

magenta	+	yellow	+	cyan	=	black

magenta	+	cyan	=	blue-violet
cyan	+	yellow	=	green
yellow	+	magenta	=	red-orange

3. OPTICAL SYNTHESIS: ORGANISM

blue-violet	+	green	+	red-orange	=	gray
yellow	+	magenta	+	cyan	=	gray

1. magenta	+	cyan	=	blue-violet	+	white
2. cyan	+	yellow	=	green	+	white
3. yellow	+	magenta	=	red-orange	+	white
4. red-orange	+	green	=	yellow	+	black
5. blue-violet	+	red-orange	=	magenta	+	black
6. green	+	blue-violet	=	cyan	+	black

The Linguistic Standpoint

A "natural" color system based on the four psychologically primordial colors (red/green/yellow/blue) survives in the antagonist system of Hering, which posits an opposition between the two pairs — red/green and yellow/blue. This conception is at odds with the three-receptors model of Young-Helmholtz. In fact both these theoretical conceptions are accurate, if it is acknowledged that for the two pairs red/blue-green and yellow/blue it is a question not of visual primaries but of complementaries whose characteristics are of maximum contrast, and which function on two exactly opposite axes. This question is treated in detail in the caption to Diagram 1 on page 60.

A study correlating color perception and linguistic data has revealed, in vocabularies of Romance, Germanic, and Slavic origin, the universal presence of names for the following colors: white, black, red, green, yellow, and blue. In languages with the most limited vocabularies the words make their appearance in the following order: white and black, denoting bright and dark; then, white, black, and red; either green or yellow is then added; afterwards two terms to distinguish between green and yellow. In the next phase blue appears; at a much later time names for the other tints are added.[3]

Why is yellow assigned the same rank as the three additive primaries red/green/blue? We must recapitulate: before the linguistic phase in which two distinct words are assigned to yellow and green, there is a word doing double duty as either yellow or green.

red — green or yellow — blue

Let us analyze the above constellation of words from the linguistic standpoint. Red comprises the whole scale running from magenta to red-orange. Thus under the name "red" are classed both the primary additive, red-orange (magenta + yellow), and the primary subtractive, magenta.

For blue, the range of choice is on the same order. All the tints ranging from cyan (primary subtractive) to blue-violet (primary additive) can be designated as blue.

Only yellow, a primary subtractive like magenta and cyan, possesses no equivalent primary additive. Likewise green, a primary additive, has nothing corresponding to it on the subtractive side. Yellow is equidistant from the red pole and the green pole of which it is composed. The sensitivity of cones to the spectrum is at its most acute in the yellows (subtractive system) and in the yellow-greens (additive system), where brilliance is highest and in consequence sharply accentuates differences of brightness and hue.

Conversely magenta and red-orange, while of very dissimilar chromatic quality, are very close in terms of brightness. As for the blues, they can be distinguished from each other solely by their magenta content (i.e., cyan = 0% of magenta; blue-violet = 100% magenta).

Only yellow evokes by its terminology a precise sensation tolerating no deviation in either direction — in one case it becomes yellow-green, in the other it becomes orange. Nevertheless it is interesting to consider the fact that deep yellow (darkened either by black or by a complementary) becomes olive and thus takes on a greenish aspect.

The ambivalence of the chromatic phraseology "green or yellow" can be explained by the imprecision of the terms blue and red even at the present time. The following formulations are exact: red — green — blue, if we are talking about the primary additives, three monochromatic radiations of the spectrum, or red — yellow — blue, if we are talking about the three subtractive primaries magenta, yellow, and cyan.

The Thresholds of Vision

What are the principal colors, corresponding to the mechanism of sight? The absorption curves of maximum spectral sensitivity gravitate around three centers: short wavelengths (blue-violet), medium wavelengths (green), and long wavelengths (red). Thomas Young was among the first to seek the explanation of the three basic colors not in the nature of light but in the human constitution. Helmholtz (1821–1894) developed this idea further. His trichromatic principles are often used in the elaboration of present-day models, whether in confrontation or in combination with the antagonist theory of Hering.

The three additive base colors, corresponding to short (blue-violet), medium (green), and long (red) wavelengths, are complemented by the three subtractive base colors (equivalent to yellow, cyan, and magenta inks and filters). To consider only the three additive base colors as primaries of the eye is the logical consequence of an exclusive reliance by the optical researcher on spectral luminosities.

On the other hand, to declare that the only primaries of the eye are the three subtractive primaries from which all the nuances of matter can be reconstituted would be similarly inconsistent. (This was the subject of a famous dispute between Goethe and Newton.)

In fact these are just two perspectives on the same problem, or rather two extremes of investigatory method which can be reconciled beautifully by the laws of complementarity and simultaneity. This subject is treated in greater depth in the chapter "Successive Image."

On one side there is the reality of light rays, on the other the exigencies of matter — but our visual apparatus needs energy as much as it needs matter in order to convert the stimuli coming from the external world into impressions of color, with their multiple intellectual and emotional repercussions.

Harald Küppers[4] suggests a fourth primary — black for the additive system and white for the subtractive system — given the fact that luminous values stand out against a black background whereas separate transparent layers of ink require a white background. White and black are the two limiting points of reference for the neutral values, the grays, which can be realized, with the addition of 40% to 80% black, from pairs of optical complementaries.

In the following chapters we will see that black as well as white can be created solely from saturated colors, not only complementaries, and by darkening as well as by brightening, through optical mixing, the six basic colors, which here represent the six fundamental thresholds of vision.

It must be stressed that we are talking about six theoretical, idealized values; at the present time it would be as impossible to determine absolutely the three primary inks that would reconstitute the perceptible richness of all nuances of matter as it would be to try to specify exactly which three wavelengths would generate the whole spectrum.

It is important that the three fundamental spectral colors differ from each other as much as possible. Here for example are several lots of colors, measured in millimicrons, which neutralize into white when they are combined:

Blue	Green	Red	
435, 8	546, 1	700	C.I.E. (International electric commission)
470	535	610	color television (F.C.C.)
465	555	632	color film (Ralph Evans)
452	538	640	color reproduction (Schumacher)
445	520	620	physiology (Wald and Brown)

For instance, the primary additives used for the triplets of television images are not identical with the three secondary base colors of the subtractive synthesis. This is particularly true of the primary additive emitted by the blue block of the triplet. The blue of 470 millimicrons has a superior luminosity to the blue-violet of 435 millimicrons. The basis for the choice is thus due to the technical impossibility at present of covering a wider field, and not to the fact that the blue in question is closer to the blue that is the optical complementary of yellow, as certain writers seem to think.

For the three primary inks (yellow, cyan, and magenta) the optimal coloring agents have yet to be found. Practically every manufacturer of ink offers his own scale, and standardization on the international level has scarcely begun. Printing has fewer tints at its disposal than photography has. In practice yellow is usually satisfying, while magenta and cyan leave something to be desired. It is magenta that is the most difficult to render accurately. If the magenta has the least tendency toward red-orange, the resulting violets will be dull; a magenta with blueish tendencies will on the contrary produce a dirty orange. (The purity of the inks used in this book can be compared with that of other inks on pages 68–69.)

Finally, if we hypothesize as fundamental physiological hues a set of limiting color impressions, we obtain six thresholds that are strictly interdependent since in a given system (additive, subtractive, or optical) the characteristics of any three of them will determine the specifics of their three complements.

The six thresholds of chromatic sensitivity correspond roughly to the six principal colors, obtained from the three primary inks: on the one hand yellow, cyan, and magenta; on the other (as this is a printed book, the three complement colors are obtained by subtraction) yellow + magenta (red-orange), yellow + cyan (green), and magenta + cyan (blue-violet).

They are analyzed here pair by pair in their complementary interaction. By optical combination we will create: blue-violet and yellow; green and magenta; red-orange and cyan.

The following rule can be stated: The secondaries red-orange, blue-violet, and green, when combined optically by pair, create hues mixed with black; the same is true for any pair containing yellow, cyan, and magenta.

On the contrary, any pair composed solely of two subtractive primaries results in a hue mixed with white.

In reality, surprisingly enough, and as we will see in the chapter on the complementaries, the six basic colors are not optically complementary among themselves. Studying the successive image will enable us to understand that this contradiction exists only in appearance.

The Simultaneous Effect

Extrapolation between objects and the space that separates them is of prime importance in perception. It is sufficient to recall those symmetrical silhouettes where the brain hesitates between two equally probable interpretations; depending on the meaning attached to the contours, background suddenly becomes foreground and vice versa (e.g., a black vase on a white background or two white facial profiles on a black background).

Attention is always directed to places in which there is activity rather than to areas of constant brightness and hue which do not elicit much inquiry.[5] The eye possesses an extraordinary power of differentiation, thanks to its faculty for surrounding a color with a simultaneous complementary halo that reverberates through all the space in the same visual field.

This dynamic mechanism allows the brain to discern infinitesimal differences in brightness, hue, and saturation, 10 million separate nuances according to an estimate made by Judd and later by McAdam.[6]

Thus two similar shades can be distinguished from each other especially along their common border by being combined with the simultaneous complementary of the other. For example, one violet, slightly redder than another with which it is juxtaposed, will accentuate its difference by the complementary green optically added to the violet with the lesser amount of red (violet + green = blue).

Positioning and quantity:

1. A color dominates another completely by its size and surrounds every part of it. The color floods the smaller surface with its simultaneous complementary, to the extent of transforming the character of the surface.

2. If there is no significant quantitative difference, there is a reciprocal simultaneous influence between a color and its setting.

3. Two colored surfaces of the same dimensions have a common border. Each color will extend across that border — the effect will be equally strong on both sides.

4. Two colors in the same immediate vicinity can have a simultaneous ascendancy without touching each other.

Diagrams 1 through 4 show only two colors. But usually a color is in a more complex situation. Many colors may have an effect that reinforces a tendency, while others may contradict it. (See the example of yellow-green on pages 128 and 129.) The nearest, largest, and most brilliant color has the most decisive simultaneous influence on another.

Brightness

In order to examine the simultaneous effect of brightness, the variables relating to hue and saturation must be eliminated. The conditions are set forth on pages 118 through 123, with white and black giving a maximum contrast of modulation to a medium gray.

Hue

The constituents of a color have a maximum impact on the transformation of its hue. (See the examples on pages 25–27, 39–41, and 53–55.) The saturated color has the most marked simultaneous ascendancy over the hue. The effect diminishes with desaturation, until the grays, whites, and blacks are reached; they are neutral, for which reason they are encroached upon and marked with the character of the complementary of any adjacent saturated hue. This unilateral impregnation of hue is most spectacular when a luminous color surrounds a gray of the same brightness. The gray will appear to be tinged with the simultaneous complementary of the background.

Saturation

The simultaneous complementary has the capacity to augment the saturation of a color to the maximum degree, without altering its hue.

On the other hand, a more saturated color of the same hue causes the maximum of desaturation. A demonstration with the six basic colors can be found in the chapter on complementaries, pages 73, 81, 89, 97, 105, and 113.

In a case in which two complementaries compose together, in the proportions of optical mixing, the surrounding background of a neutral gray, the simultaneous effects cancel each other out and the gray remains unchanged. Three examples give the proof of this on pages 16, 30, and 44 (diagram B) with the successive image of gray.

Simultaneous ascendancy does not take place in two cases:

1. The distance between two colors is too great. They are no longer in the same visual field.

2. The elements constituting a structure lose their individuality at a certain distance from the observer; they fuse into a new unity which in its turn is besieged by the influences of other unities.

Simultaneity and optical mixing are two opposite phenomena: the first accentuates differences, the second annuls them.

THE INFLUENCE OF MEMORY

Stand far enough away from the book so that the small dots can no longer be distinguished from one another. If this condition is observed, the surface made up of dots separated by about 1.5 mm. and 2 mm. will give way to an impression of uniform color. The simultaneous effect will then come into play for the two small squares in the center of the backgrounds of saturated color. Without the precaution of observing from far enough away, the comparative calculation may inhibit and contradict your subjective impression, since the size of the dots makes it possible to know which color predominates, and this knowledge may influence your judgment when defining the tint of a color. Here logic and memory intervene to such an extent that the eye no longer sees: it knows. The hypothesis formulated by the intellect will take precedence over the evidence of the senses. In order to give free rein to simultaneous influences, optical mixing by the eye is called for. The distance required for abstracting the structure will vary from reader to reader. Some experimentation is necessary in order to benefit from the full amplitude of the mechanism in question.

Unlike the case in which one guesses the appearance of a color on the basis of the size of the dots, the fact of knowing in advance what aspect a color will take, under which simultaneous complementary conditions, will aid in the rapid and accurate perception of the effects. The expectation of a particular color sensation gives one the patience to open up to the subliminal sensory stimuli which normally remain unconscious in everyday life.

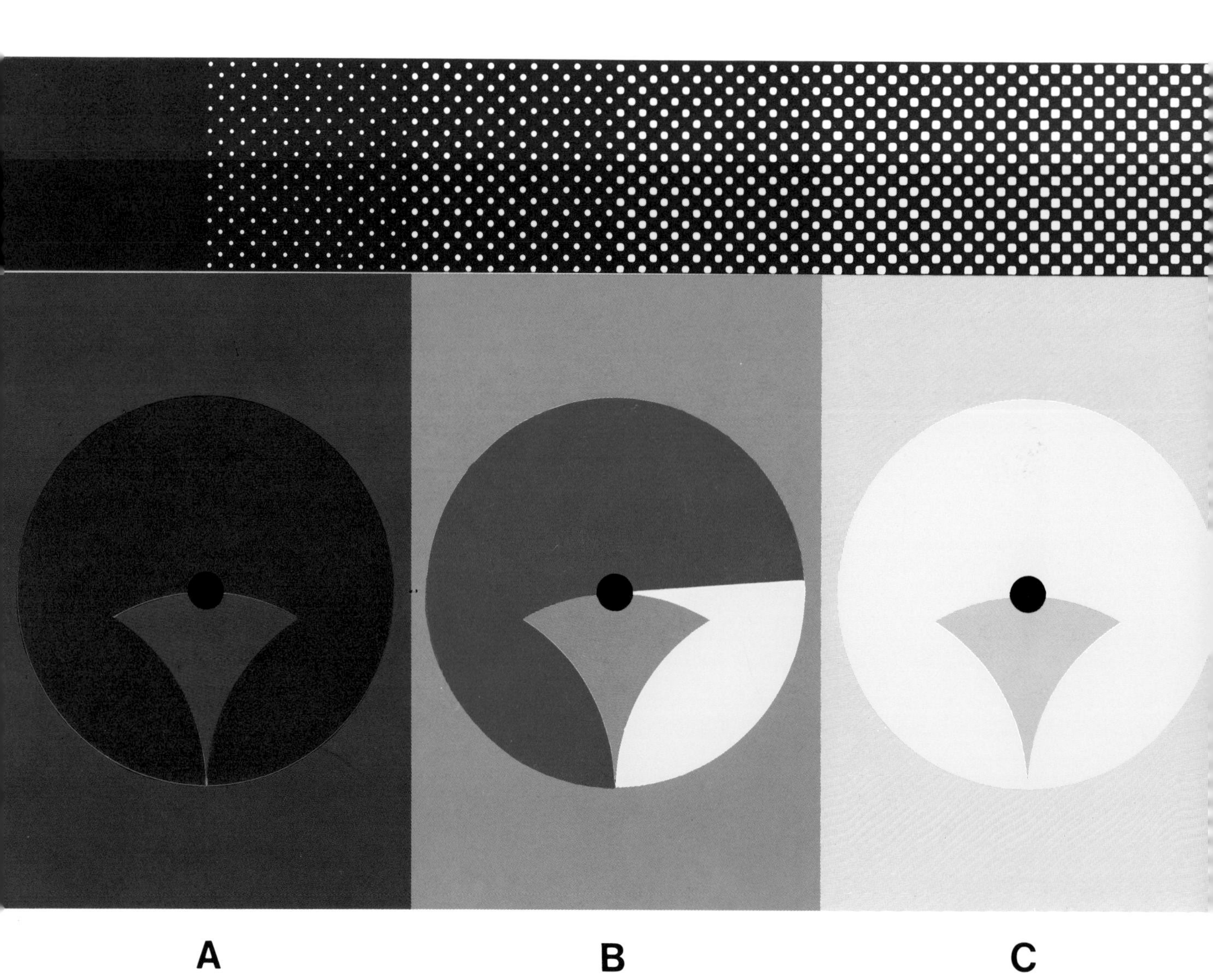
A
B
C

Blue-Violet and Yellow

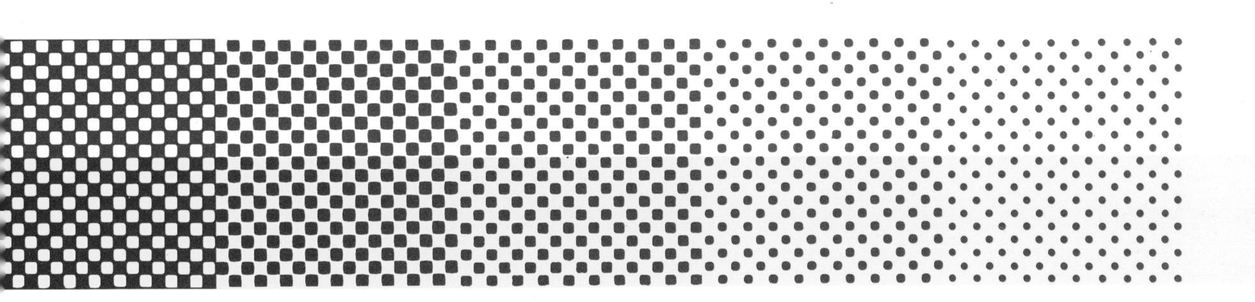

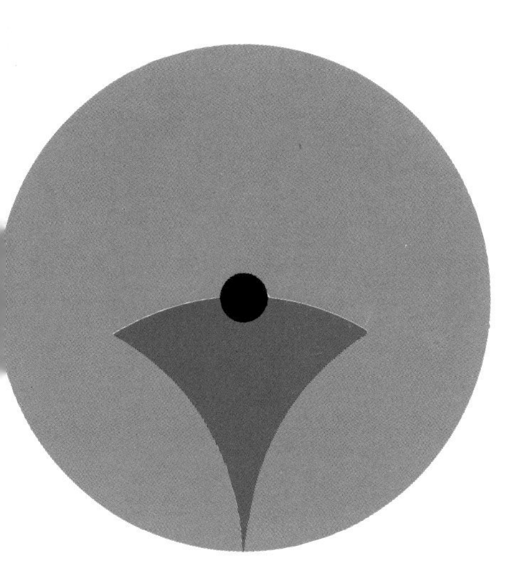

D

Yellow is the subtractive complementary of blue-violet (100% cyan + 100% magenta). As in the scale above, it is sufficient to add a yellow saturated to 100% for all the examples up to page 29. Two transparent yellow sheets are included for that purpose.

After studying the chapter on the successive image:

A. The simultaneous complementary of the blue-violet, concentrated in the curvilinear gray triangle of the same brightness, changes back into blue-violet in the successive image.

B. Yellow, with its optical complementary in the proportions of optical mixing, retains the neutrality of the triangular gray in the successive image as well.

C. In the successive image there may be observed a yellow triangle which is the complementary of the yellow's complementary. The yellow orange which is successive complementary of blue D is not more powerful than this in terms of luminosity and saturation.

From White to Blue-Violet

With the four variations on these pages, the scale ranging from white to blue-violet can be realized through optical mixing:

1. Magenta of 10%-30%-50%-70%-100% and the red-violets, optically mixed with 100% cyan.

2. Cyan of 10%-30%-50%-70%-100% and the blues, mixed optically with 100% magenta in the above proportions.

3. Blue-violet + white in the above proportions.

4. Magenta of 10%-30%-50%-70%-100% and the red-violets (pages 22, 23 scale a) optically mixed with cyan of 10%-30%-50%-70%-100% and the blues (scale c) in the ratio 45%:55% (scale b on pages 22 and 23).

45

55

1

45

55

2

3

From Yellow to Black

By superimposing the transparent yellow sheets on the four variations on these pages, the scale from yellow to black can be created through optical synthesis:

1. The oranges and red-orange + black of 10%-30%-50%-70%-100% optically mixed with 100% green in the above proportions.

2. The yellow-greens and the greens + black of 10%-30%-50%-70%-100% optically mixed with 100% red-orange in the above proportions.

3. Yellow + black in the above proportions.

4. The oranges and red-orange + black optically combined with the yellow-greens and the greens + black in the ratio 45%:55%. (See scales a, b, c on pages 22 and 23.)

45

55

4

Blue-Violet

55 parts cyan + 45 parts magenta = blue-violet + 50% white, halfway between white and 100% blue-violet.

Disk a indicates the above proportions. In rotation the magenta and cyan combine optically into the blue-violet + white of disk b, which is the average in hue, brightness, and saturation of disk a. This same medium blue-violet can be found on page 19 on scale 3 (see arrow), with 30 parts white and 70 parts blue-violet.

The red-violets and the blues desaturated by white:

By successively increasing the parts of magenta to 50, 60, 70, 80, and 90 parts we obtain the scale of red-violets below. Conversely, by increasing the part of cyan to 60, 70, 80, 90, and 95 parts we obtain the blues.

Desaturation by white attains its culminating point in blue-violet (see arrow).

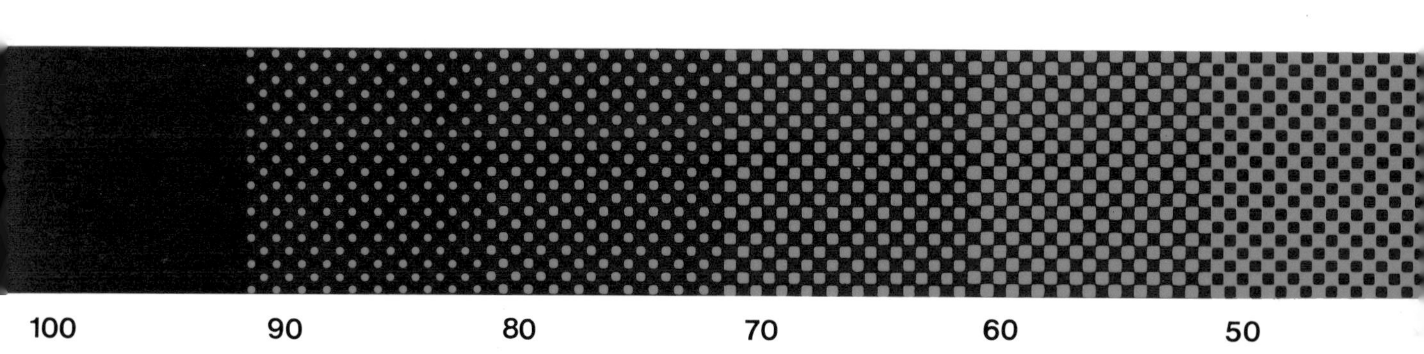

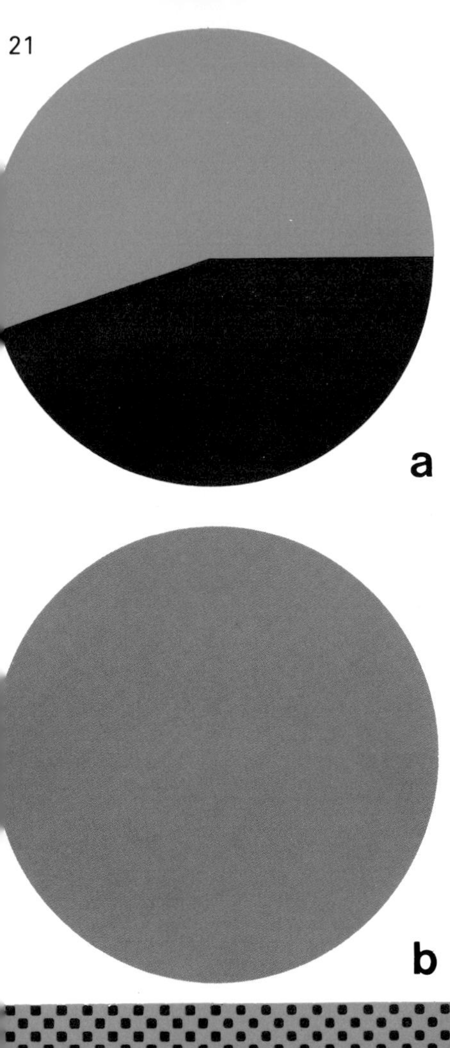

+ *Yellow*

55 parts green + 45 parts red-orange = yellow + 50% black, halfway between black and 100% yellow.

Red-orange and green have a similar brightness; consequently approximately the same brightness can be observed in the 12 stages of the scale below. As the oranges and yellow-greens gain in yellow, the percentage of black progressively augments to culminate in a neutral yellow with about 50% black added to it (disk b).

The oranges and the yellow-greens desaturated by black:

When the red-orange exceeds 45%, all the desaturated oranges are found to the left of the arrow.

The desaturated yellow-greens to the right of the arrow begin at 55% green.

40 30 20 10 5 0

B

Successive Image:

In successive inversion on a black background you will find the same colors as are obtained by adding a transparent yellow circle, except that the two semicircles are reversed in the successive image, the orange being on the left and the green on the right.

The left-hand disk + one yellow circle: the successive image gives you the original disk with the semicircles reversed and without yellow. On a gray background, the successive image gives the simultaneous complementaries.

+ *Yellow*

Add the two transparent yellow sheets: scale a is transformed into oranges and into red-orange + black; scale c into yellow-greens and into green + black. The optical mixing of a + c is a yellow scale desaturated by blue-violet complementary (b). By adding the narrow yellow stripes only to scale b, you can more easily note the optical yellow, synthesis of the oranges and the greens.

Simultaneous Effects: Blue-Violet

A red-violet and a blue both appear blue-violet.
From far enough away the two small blue-violet squares below resemble in their tint and their saturation the two squares in the center of the magenta background and in the center of the cyan background.

The red-violet surrounded by magenta seems more violet, while the blue surrounded by cyan brings out the magenta elements. There is a minimum contrast of hue and a slight decrease in saturation.

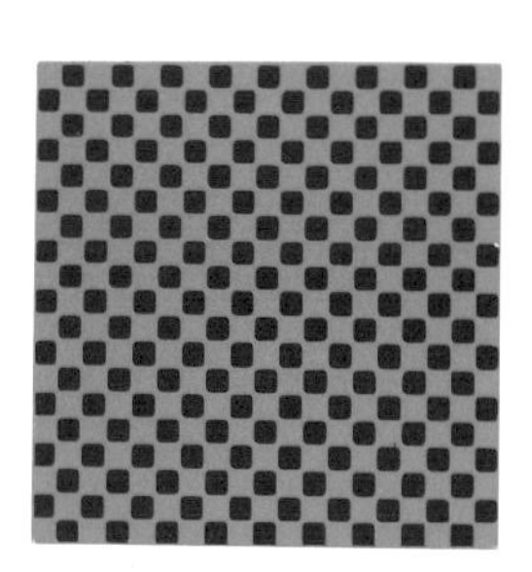

+ *Yellow*

After superimposing the transparent yellow squares on the two blue-violet squares on the opposite page, and a large yellow rectangle on the squares below, you will observe that the two dark yellow squares on page 24 resemble the two squares in the center of the respective red-orange and green backgrounds in both hue and saturation.

A brown and an olive both give an impression of deep yellow.

The brown surrounded by red-orange appears duller and greener than in reality. The olive tone surrounded by green becomes somewhat more desaturated and gives a yellower impression.

Before turning the page, put the transparent yellow sheet on the bottom of page 25 next to the large rectangle on page 27; if you do not look until it is in place, the surprise will be all the greater when you compare the astonishing contrast effects due to the inversion of the backgrounds.

45 parts of magenta dots + 55 parts of cyan dots become indistinguishable against a uniform background of blue-violet of 50%.

Opposite page:
Maximum contrast of hue and augmentation of saturation.

Here the simultaneous effect on page 25 is reversed. This time the cyan background emphasizes the brightness of the magenta dots and the surrounding magenta activates the brilliance and prominence of the blue hue.

By adding the white yellow strip on page 26, the 50% of blue-violet becomes 50% black. The dark yellow background thus created is indistinguishable from the structure, which, when seen from far enough away, optically combines 45 parts red-orange and 55 parts green.

In the example on page 22, the dark yellow is surrounded by equal parts of green on one side and red-orange on the other. It does not undergo any change, because the green and the red-orange emit simultaneous chromatic impulses that are contradictory and thus cancel each other out in the yellow. Nevertheless a cleavage is visible, with a green tendency along the red border and a red fluctuation on the green border. The seeing eye, unable to decide for either red or green, reconstitutes the medium optical yellow.

Maximum contrast of hue and augmentation of saturation.

Cyan, complementary of the red-orange background, when combined optically with the small red-orange elements of the green, emphasizes the green components of the color olive (olive + cyan = green), while the green frame draws the eye to the red elements of the central brown (magenta + orange = red).

A blue-violet appears red-violet or blue.

The subtractive generators of blue-violet are the pair with the most pronounced impact on hue. The complementary of magenta is green (blue-violet + green = cyan), hence magenta brings out a blue tendency.

The complementary of cyan is orange (blue-violet + orange = magenta), which means that cyan stimulates the red-violet tendency. The coefficient of simultaneous influence corresponds to the quantitative difference between the magenta and cyan of the blue and red-violet on the opposite page.

A yellow appears orange or green.

The red and green which make up deep yellow have the most striking simultaneous impact because magenta, complementary of green, combines optically with yellow (yellow + magenta = orange) and cyan which is complementary of red-orange draws out the green component of the yellow (cyan + yellow = green).

Mention must also be made of the fact that magenta and cyan can cause a cleavage of yellow into red and green. The violets and the blues, on the other hand, barely influence hue at all. Blue-violet and blue, respectively the subtractive complementary and the optical complementary of yellow, augment its brilliance and its saturation to the highest degree.

The coefficient of influence of the simultaneous effects on the opposite page corresponds to the quantitative difference between the two colors below:

For blue-violet:

60% magenta + 40% cyan: the red-violet square A at left.

30% magenta + 70% cyan: the blue square B at right. The difference between A and B is of 30%.

For yellow:

60 parts red-orange + 40 parts green = brown A at left.

The olive at right is composed of: 30 parts red-orange + 70 parts green. The difference in question is of 30%.

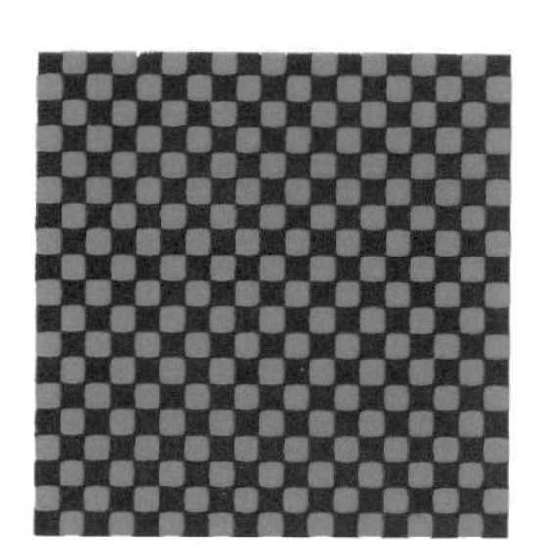

A

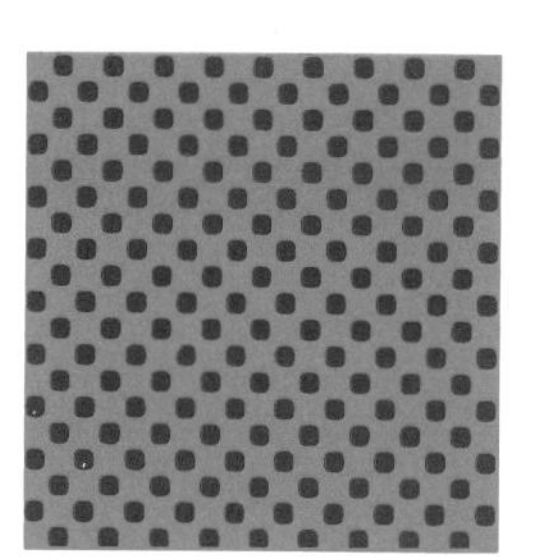

B

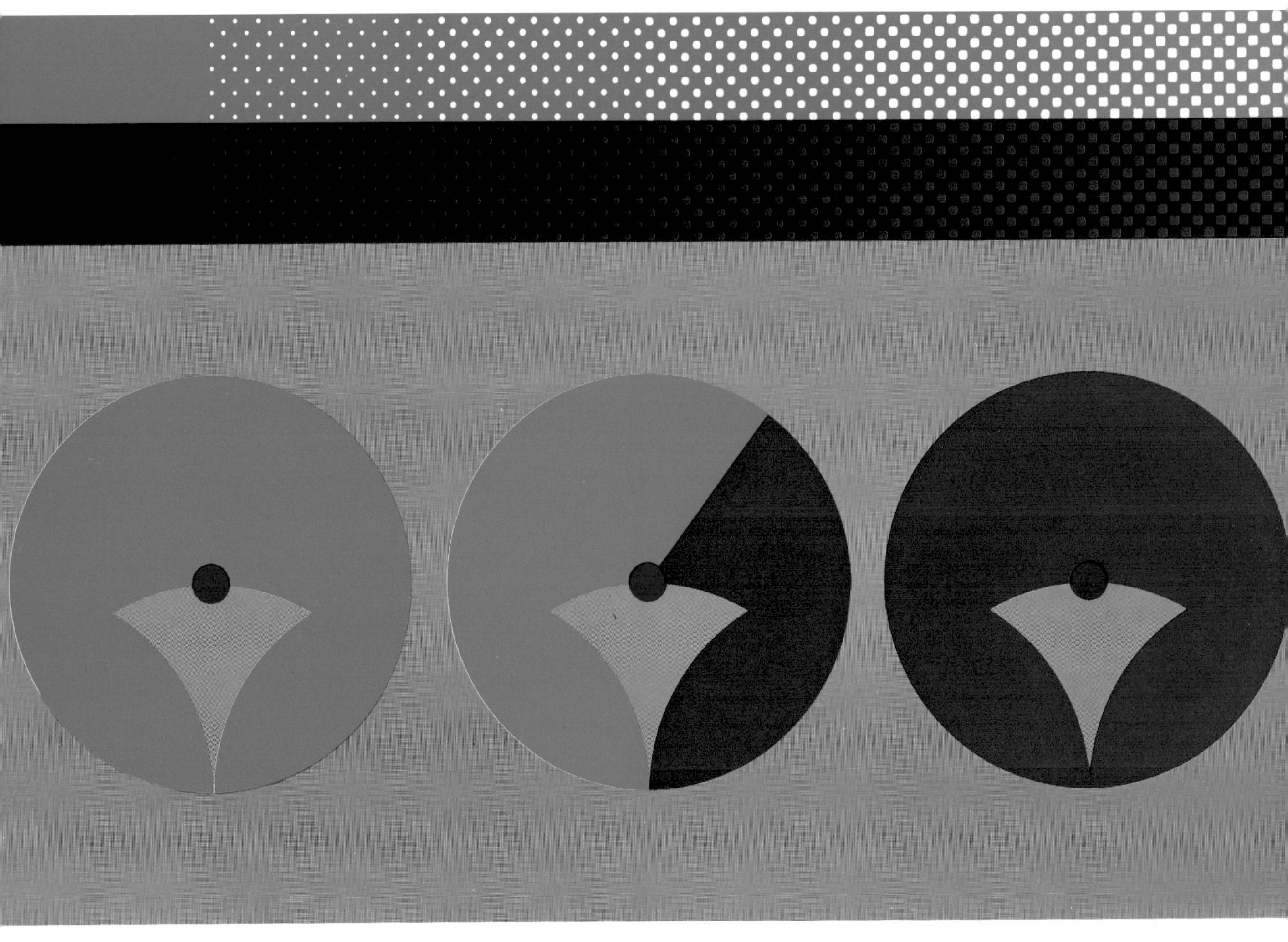

A B C

Green and Magenta

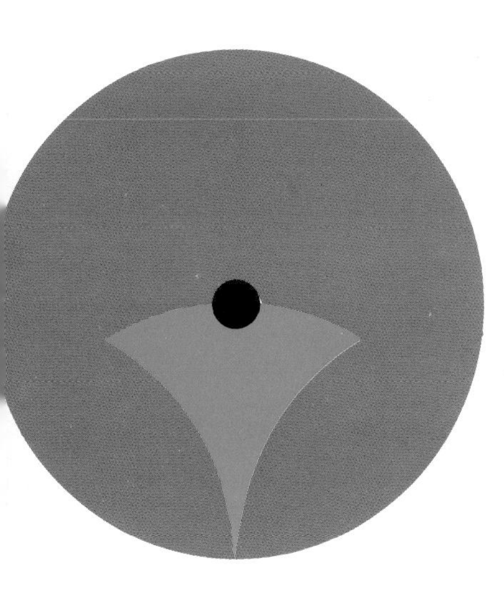

D

Magenta is the subtractive complementary of green (100% cyan + 100% yellow).

As in the scale above, it is sufficient to add a magenta saturated to 100% for all the examples up to page 43. Two transparent magenta sheets are included for that purpose in a special pocket.

After studying the chapter on the successive image:

A. The simultaneous complementary of green, concentrated in the curvilinear gray triangle of the same brightness, changes back into green in the successive image.

B. The magenta with its optical complementary, green (100% cyan + only 70% yellow), in the proportions of optical mixing retains the neutrality of the triangular gray in the successive image as well.

C. In the successive image a magenta triangle can be perceived which is the complementary of the complementary of magenta. The magenta which is the successive complementary of green D is not more powerful than this in luminosity or saturation.

From White to Green

Optical synthesis offers four solutions for obtaining a scale ranging from white to green:

1. Yellow of 10%-30%-50%-70%-100% and yellow-green mixed optically with 100% cyan in the above proportions.

2. Cyan of 10%-30%-50%-70%-100% and blue-green mixed optically with 100% yellow in the above proportions.

3. Green + white in the above proportions.

4. Yellow of 10%-30%-50%-70%-100% and yellow-green + cyan of 10%-30%-50%-70%-100% and blue-greens combined optically in the ratio 40%:60%. (See scale b, page 36 and 37.)

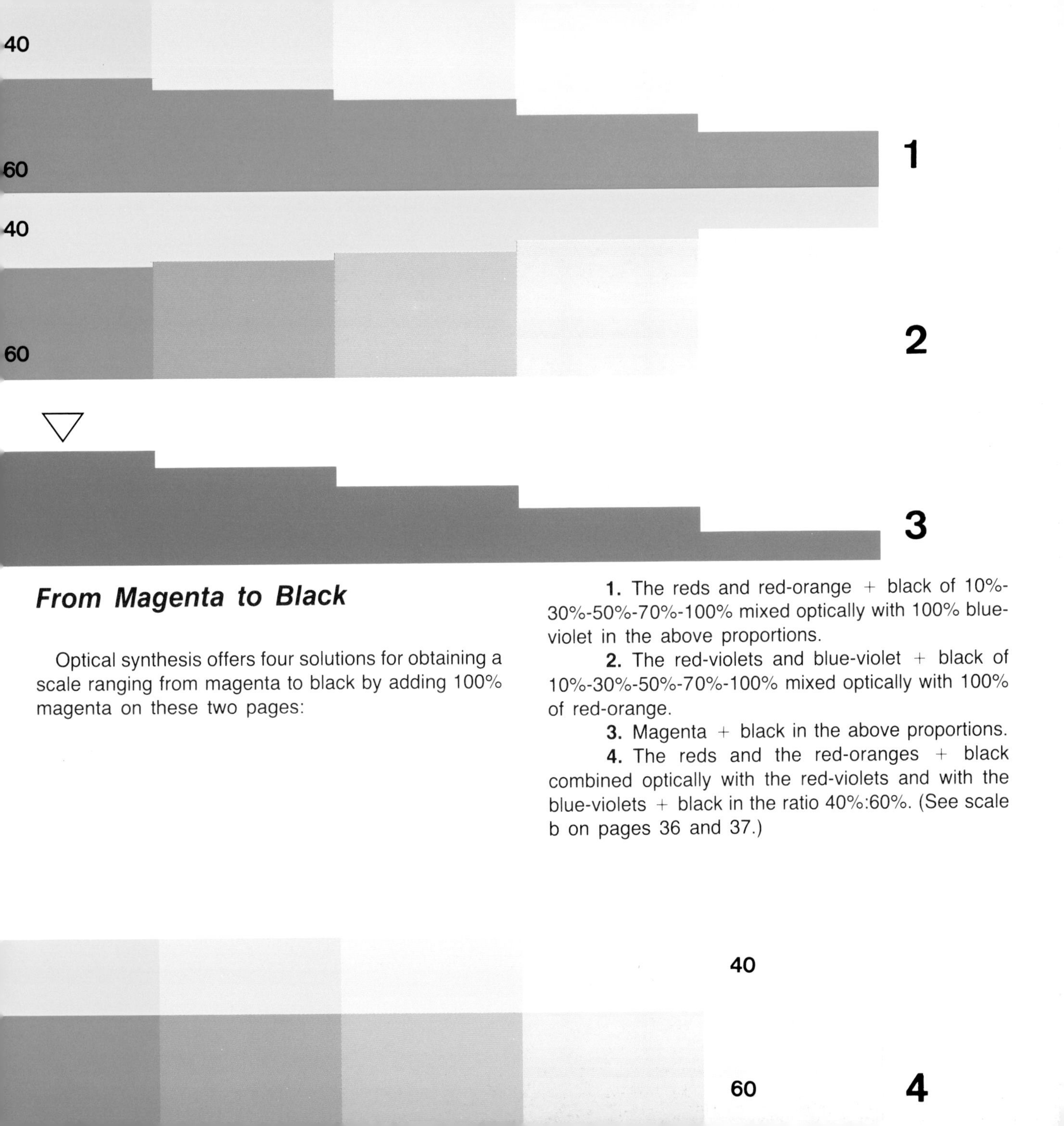

From Magenta to Black

Optical synthesis offers four solutions for obtaining a scale ranging from magenta to black by adding 100% magenta on these two pages:

1. The reds and red-orange + black of 10%-30%-50%-70%-100% mixed optically with 100% blue-violet in the above proportions.

2. The red-violets and blue-violet + black of 10%-30%-50%-70%-100% mixed optically with 100% of red-orange.

3. Magenta + black in the above proportions.

4. The reds and the red-oranges + black combined optically with the red-violets and with the blue-violets + black in the ratio 40%:60%. (See scale b on pages 36 and 37.)

Green

Yellow is a very bright color which corresponds to about 10% black; cyan is of medium brightness and corresponds to about 50% black. The green which results from the optical mixing will have an intermediate brightness, comparable to a gray of about 30% black. Disk b, with the transparent magenta of 100%, shows clearly a deep magenta with a black component of 30%. The purity of the cyan ink determines how clean the optical green will be; even a minute quantity of magenta in the cyan can cause a gray fogging effect.

60 parts cyan + 40 parts yellow = green + 50% white.
Disk b shows a green of 50% which represents an optical mixing of the yellow and cyan sections of disk a.

The yellow-greens desaturated by white.
Every nuance of yellow-green brightened by white between green and yellow, obtained by optical combination, is located between 0% cyan (100% yellow) and 60% cyan (40% yellow).

The blue-greens desaturated by white.
The blue-greens + white between green (see arrow) and cyan are optical combinations with the yellow being of 0% to 40% and the cyan of 60% to 100%.

The red-violets desaturated by black.
The red-violets situated between magenta + 30% black and saturated blue-violet require between 60% (40% red-orange) and 100% blue-violet (0% red-orange).

100 95 90 80 70 60

A

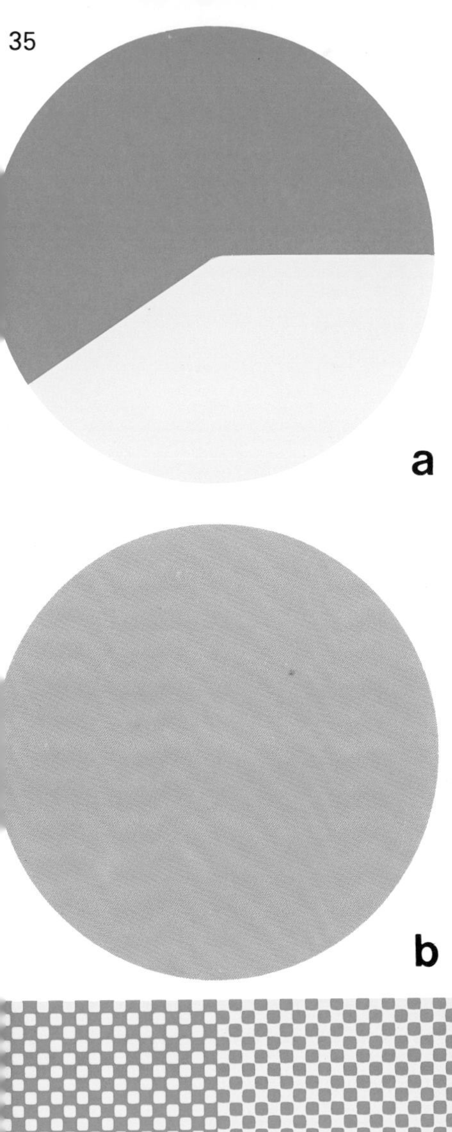

+ *Magenta*

60 parts blue-violet + 40 parts red-orange = magenta + 30% black.
Simply by adding the two magenta circles to disks a and b, we keep the same proportions as for the optical synthesis of green. Disk b: 50% green changes into 30% black.

The reds desaturated by black.
The reds between saturated red-orange and magenta + 30% black are obtained by optical mixing with a blue-violet component ranging from 0% (100% red-orange) to 60% (40% red-orange).

50 40 30 20 10 0

△

B

Successive Image:

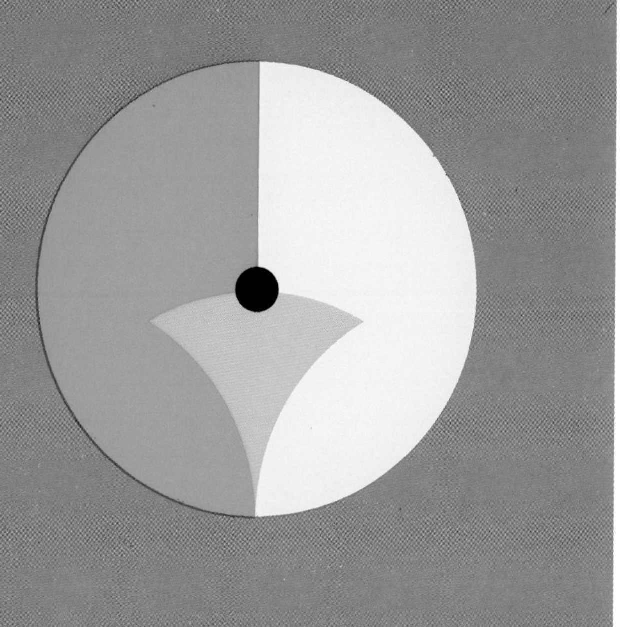

In the successive image on a black background the same colors will be found as by adding the transparent magenta circle, except that the semicircles are inverted, with blue-violet on the right and orange on the left.

The left-hand disk + one magenta circle: The successive image yields the original disk, with the semicircles reversed and without magenta.

On a gray background the successive images tend more toward the simultaneous complementaries.

+ *Magenta*

Add the two transparent magenta sheets: scale a changes into reds and into red-orange + black, while scale c changes into red-violets and into blue-violets + black. The optical mix of a + c is a magenta scale desaturated by the green complementary.

By adding the narrow magenta strips to scale b, you can more easily examine the optical magenta, a synthesis of the reds with the violets.

a

b

c

Simultaneous Effects: Green

From far enough away, the two small green squares resemble the two colors in the center of the yellow background on the one hand and in the center of the cyan background on the other.

Minimum contrast of hue and diminution of saturation.

A yellow-green and a blue-green give the impression of a green of the same brightness and the same hue. The complementary of the surrounding yellow gives a greener, duller character to the yellow-green, while the complementary of the cyan background is responsible for the transformation of the blue-green into de-saturated green.

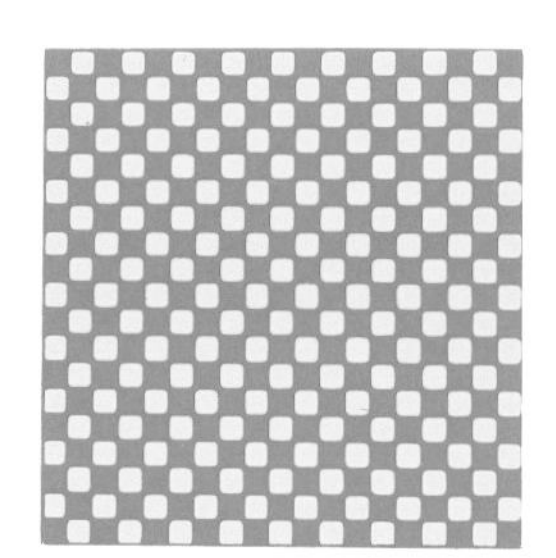

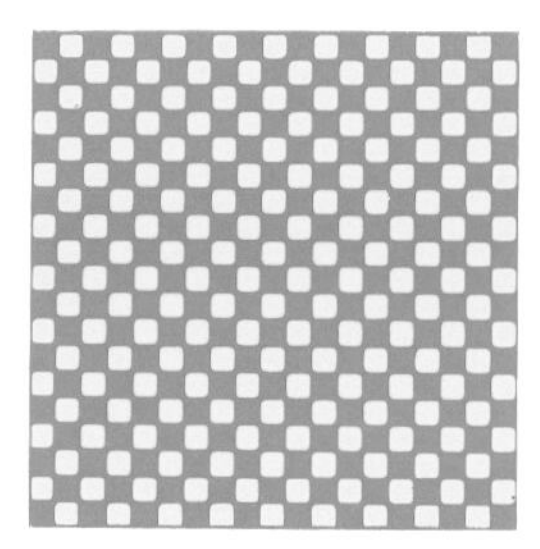

+ *Magenta*

The two squares on the opposite page resemble the two squares in the center of the blue-violet and the red-orange.

A red and a red-violet both resemble a deep magenta.

The maximal contrast of hue is accompanied by diminished saturation.

The blue-violet background brings out the red-orange elements even though they are a minority. Inversely the red-orange background emphasizes the blue-violet components.

Before turning the page, place the magenta sheet of page 39 on page 41; if you do this without watching, you will be all the more surprised when you compare the extraordinary contrast effects due to the inversion of the backgrounds.

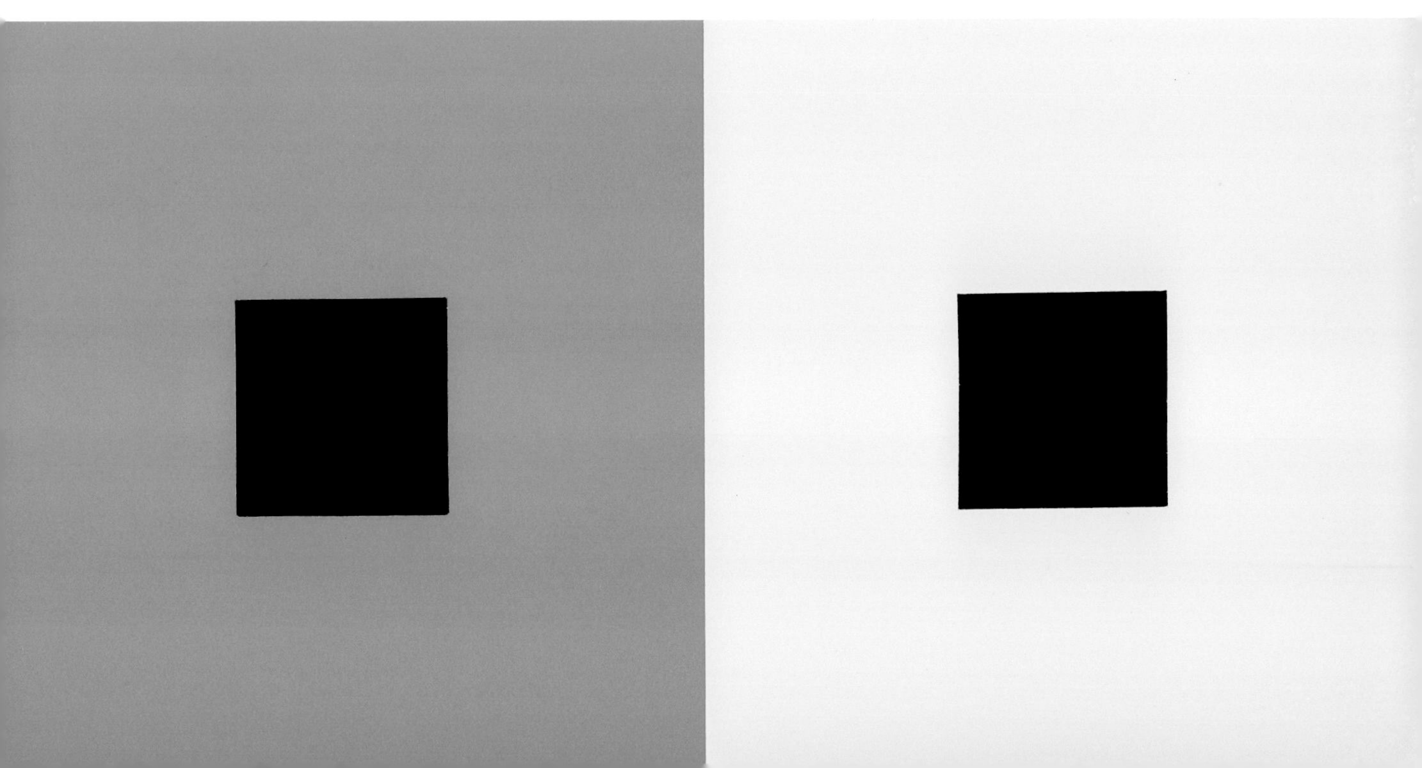

When the small saturated yellow elements are associated with the small saturated cyan elements, the hue created is close to a homogeneous green of 50% (primary green: 50% cyan + 50% yellow).

+ *Magenta*

The red-orange dots are juxtaposed with the blue-violet dots. Their structures become more or less indistinguishable against a homogeneous background of 100% magenta + 30% black.

Maximum contrast of hue and augmentation of saturation.

The effects on page 39 are reversed. The yellow-green in the center of the cyan evokes yellow; the blue-green, conversely, is pushed in the direction of cyan by the surrounding yellow.

+ *Magenta*

What appeared to be a deep magenta becomes a red-orange surrounded by blue-violet. It is the simultaneous yellow, the complementary of blue-violet, which induces in the red a tendency toward red-orange. The surrounding red-orange on the contrary has caused the deep magenta of page 39 to undergo a mutation toward blue-violet. The simultaneous cyan, complementary of the red-orange, pushes it in the direction of blue-violet by combining with the central blue-violet.

Green

A green seems either yellow-green or blue-green.

The coefficient of the simultaneous influences in play corresponds to the quantitative difference between the two colors on the opposite page. The orange, simultaneous complementary of the cyan, forms with the central green a brighter and yellower color.

The blue, simultaneous complementary of the yellow, floods the green in the center with the yellow background and thus makes it deeper and bluer. The cleavage of green into its two components, yellow and cyan, is most striking in the confrontation of a yellow background opposed to a cyan background.

Blue-green A on the left is made up of 30 parts yellow + 70 parts cyan. Yellow-green B on the right is made up of 70 parts yellow + 30 parts cyan. The quantitative difference between these two colors is of 40%.

+ *Magenta*

A magenta appears either red-violet or red-orange.

Deep magenta undergoes the greatest transformation of hue through the simultaneous contrasts of its components: red-orange and blue-violet. Next come the oranges and the blues, followed by yellow and cyan. The greens belong to the scale of complementaries that have only a small influence on hue. Simultaneous green gives magenta a maximum brilliance.

Centered in blue-violet, deep magenta resembles red B. (below right). Alongside, centered in red-orange, the same deep magenta becomes comparable to red-violet A below.

A, on the left, is made up of 30% red-orange + 70% blue-violet. B, on the right, is made up of 70% red-orange + 30% blue-violet. The quantitative difference is of 40%.

A

B

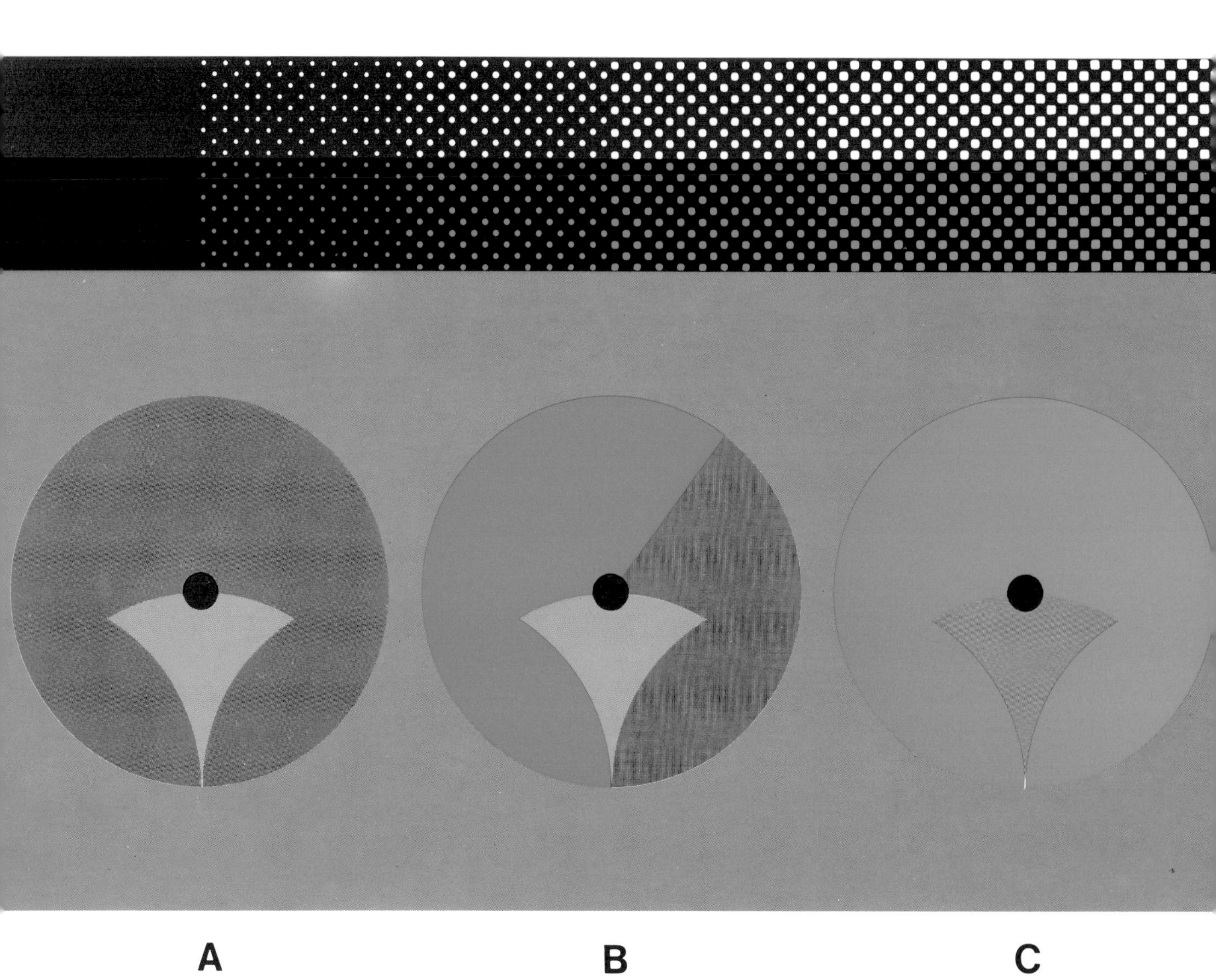

A B C

Red-Orange and Cyan

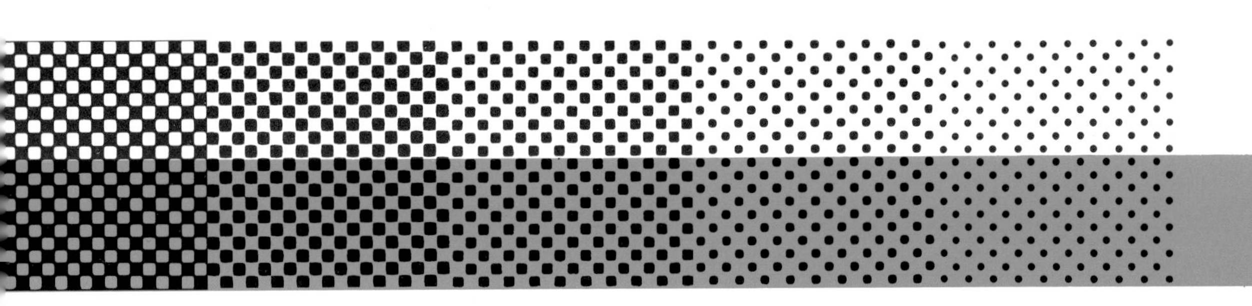

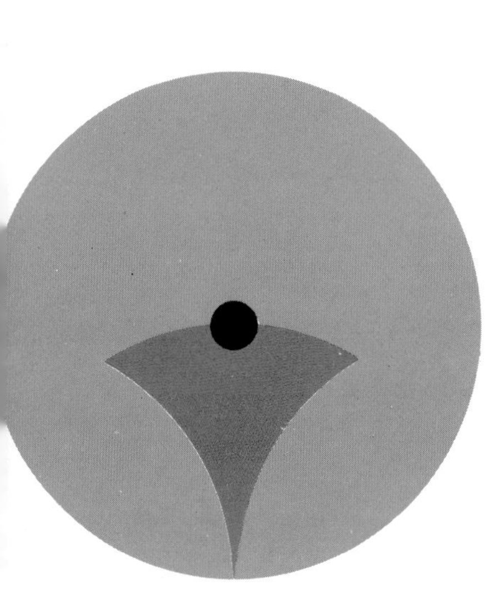

D

Cyan is the subtractive complementary of red-orange.

As in the scale above, it is sufficient to add a cyan saturated to 100% for all the examples up to page 57. Two transparent sheets of cyan are included for that purpose in a special pocket.

After studying the chapter on the successive image:

A. The simultaneous complementary of red-orange, concentrated in the curvilinear gray triangle of the same brightness, changes back into red-orange in the successive image.

B. Cyan with its simultaneous complementary in the proportions of optical mixing retains the neutrality of the triangular gray in the successive image as well.

C. In the successive image a cyan triangle can be discerned which is the complementary of the complementary of cyan. The cyan, which is successive complementary of orange D, is not more powerful than this in luminosity and saturation.

From White to Red-Orange

Using the four variations on these pages, the scale running from white to red-orange can be realized by optical synthesis:

1. Yellow of 10%-30%-50%-70%-100% and the oranges mixed optically with 100% magenta in the above proportions.

2. Magenta of 10%-30%-50%-70%-100% and the reds mixed optically with 100% of yellow in the above proportions.

3. Red-orange + white in the above proportions.

4. Yellow of 10%-30%-50%-70%-100% and the oranges mixed optically with magenta of 10%-30%-50%-70%-100% and the reds in the ratio of 35%:65%. The dots of scale b on pages 50 and 51 are divided up according to the same quantitative ratio.

35

65

1

35

65

2

3

From Cyan to Black

With the four variations on these pages, by superimposing the transparent 100% cyan sheets, you can create the scale from cyan to black through optical synthesis:

1. The blue-greens and green + black of 10%-30%-50%-70%-100% mixed optically with 100% of blue-violet in the above proportions.

2. The blues and blue-violet + black of 10%-30%-50%-70%-100% mixed optically with 100% green in the above proportions.

3. Cyan + black in the above proportions.

4. 35% of yellows and greens + black are optically combined with dots (scale b, pages 50–51) of 65% of magentas and blue-violets + black.

35

65

4

Red-Orange

65 parts magenta + 35 parts yellow = 50% red-orange.

Red-orange, a color midway between magenta and yellow, requires for optical synthesis only 35% yellow as compared with 65% magenta. The 50% white contained in the hue of disk b makes up for the difference in brightness between the brightest primary (yellow) and a primary of medium brightness (magenta).

The oranges desaturated by white.

All the hues of orange between yellow and red-orange, obtained by optical mixing of yellow with magenta, are found between 35% and 100% yellow.

The reds desaturated by white.

All the reds between red-orange and magenta resulting from the optical mixing of yellow with magenta contain less than 35% yellow.

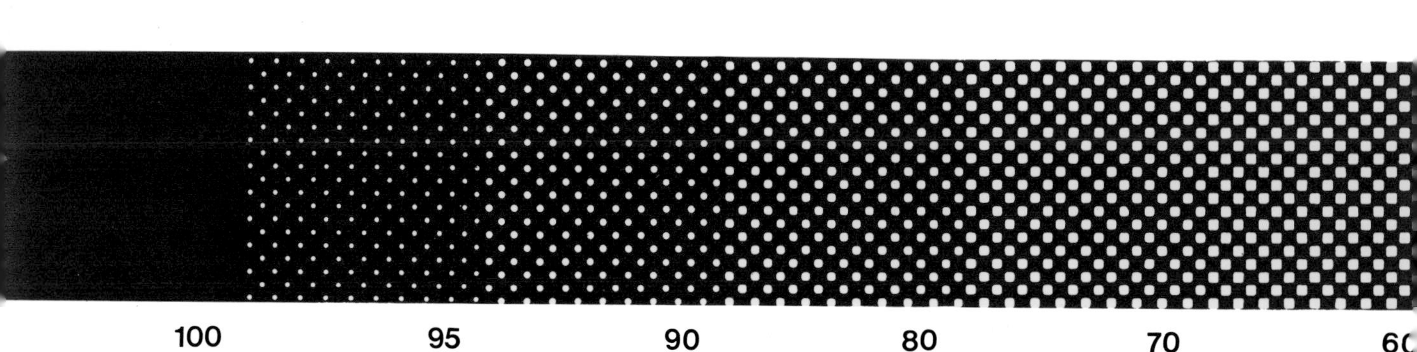

A

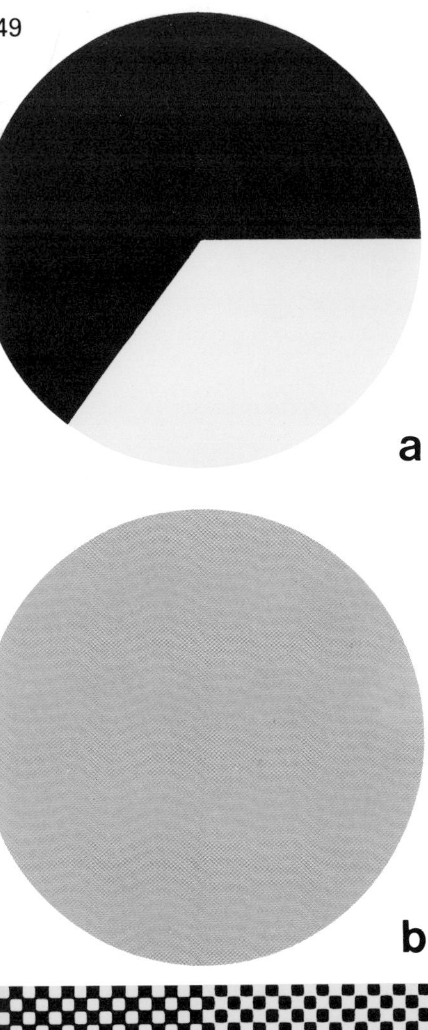

+ *Cyan*

65 parts blue-violet + 35 parts green = 100% cyan + 40% black.

Disk a reveals the above proportions, simply by the addition of the cyan sheet. Rotation of the disk results in the average hue and brightness of a deep cyan, represented by disk b. Black renders the optical mixing of yellow with magenta contained in blue-violet and green.

The blue-greens desaturated by black.

All the blue-greens which are an optical mix of the blues with the greens possess more than 35% green.

The blues desaturated by black.

The blues ranging from cyan to blue-violet are an optical mix of blue-violet with a maximum of 35% green.

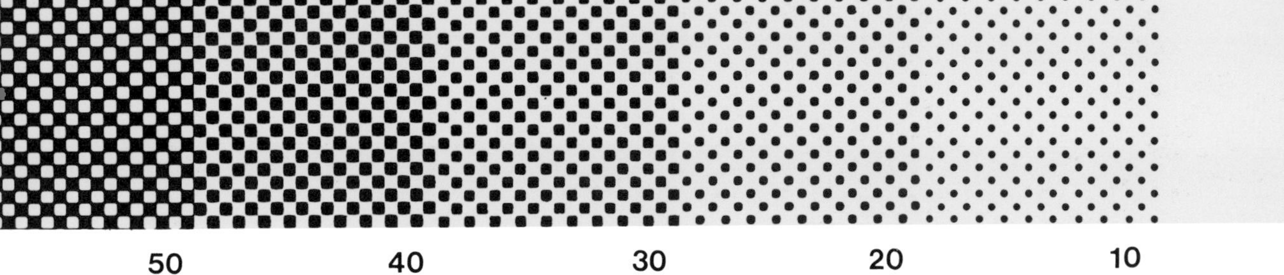

B

Successive Image:

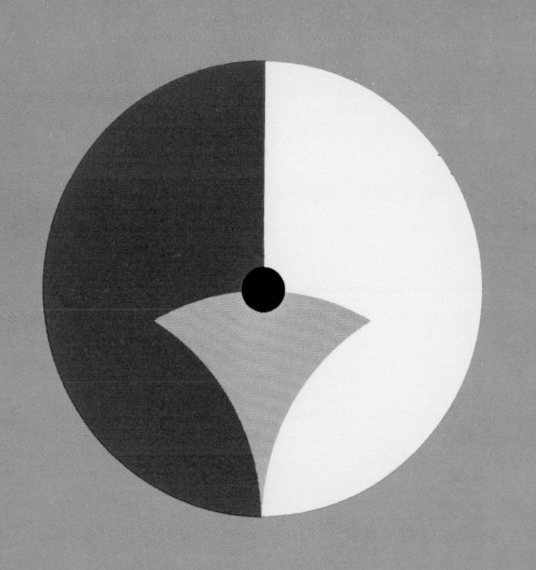

On a black background the successive image of the disk on the left yields the same colors as are obtained by adding a transparent cyan circle, except that the semicircles are reversed, with the blue-violet on the right and the green on the left.

+ 1 cyan circle:

The successive image yields the disk on the left, with the semicircles reversed.

On a gray background the successive images tend more toward the simultaneous complementaries.

+ *Cyan*

Add the two transparent cyan sheets: scale a changes into blue-greens and into greens + black; scale c changes into blues and into blue-violets + black. The optical mixing of a + c is a cyan scale desaturated by its red-orange complementary.

By adding the two narrow cyan strips to scale b, you will be enabled to take note of optical cyan, a synthesis of the greens with the blues.

Simultaneous Effects:

Red-Orange

The desaturated red-orange below resembles the two squares in the center of the yellow and in the center of the magenta on the opposite page.

Maximum desaturation.

A red and an oranged yellow appear identical.

The blue, simultaneous complementary of the yellow, draws out the magenta elements from the orange in the center at right (blue-violet + red-orange = magenta). The similarity of hue is accompanied by a similarity of brightness. The yellow-orange suggests a redder and deeper color than in reality.

The simultaneous green of the magenta background desaturates the magenta elements contained in the central red on the left and accentuates the brilliance of the yellow (green + red-orange = yellow) — that is why the red gives a slightly whiter and more orange impression. (Compare this with the white which makes up for the difference in brightness between yellow and magenta in optical synthesis.)

+ *Cyan*

Adding the cyan sheets to these two pages, the two deep cyan squares on the opposite page resemble the two squares in the center of the blue-violet and in the center of the green.

A blue and a green both appear deep cyan.

Similarity of hue and diminution of saturation of the blue at left and the green at right: The yellow-green, the simultaneous complementary (page 105), combines optically with blue, which takes on the aspect of deep cyan.

The blueish magenta, simultaneous complementary of the green background, pushes the central green in the direction of blue.

Before turning the page, add the cyan sheet from page 53 (large rectangle) to page 55 without watching; in this way the surprise will be greater when you compare the contrast effects caused by the inversion of the backgrounds.

The rapid rotation of two colors on a disk theoretically allows a synthesis comparable to that obtained by the juxtaposition of tiny strips of color in space. In one case, the duration of the visual stimulation is too short for a separate identification of the colors; in the other case, the spatial areas are too small to be perceived individually. Nonetheless, when the contrast of brilliance is considerable (yellow + cyan — yellow + magenta — yellow + blue, white + black), the optical mixing obtained by rotation is more vigorous than that achieved with dots of the dimension used in this book. Since the inertia of the eye is less when using the dots, the quantity of the brightest dots will be somewhat augmented.

The yellow and magenta dots become indistinguishable with the background below, consisting of 50% red-orange.

+ *Cyan*

A background of cyan + 40% black (corresponding to 50% red-orange) gives the same impression as a structure consisting of green dots with blue-violet dots.

Maximum contrast of hue and brightness and augmentation of saturation.

For the two small squares in the center below, an increased luminosity becomes evident, with a greater difference in hue and brightness. The effects on page 53 are reversed. The orange becomes yellower and brighter when surrounded by magenta, while a surrounding yellow causes a mutation into a redder and deeper color.

+ *Cyan*

The simultaneous effects on page 53 are reversed. This time the blue-violet elements predominant in the blue are maintained in their brilliance by the complementary influences of the green. The blue-violet background on the other hand activates the green structure.

A red-orange appears red or yellow-orange.

The red-orange juxtaposed with the green which is simultaneous complementary of the magenta background becomes yellower.

The red-orange juxtaposed with the blue-violet which is complementary of the yellow background becomes redder.

The 30% coefficient of the simultaneous influence corresponds to the difference between the two colors A and B on the opposite page.

Page 57:

30% yellow + 70% magenta = the red on the left (A).

60% yellow + 40% magenta = the orange-yellow on the right (B).

A deep cyan appears blue or blue-green.

Deep cyan's most significant transformation of hue occurs through the simultaneous effects of its optical constituents: green and blue-violet. At left they bring about a blue surrounded by a green background; at right, a blue-green in the center of the blue-violet background, comparable to the two colors on page 57.

Under the influence both of the yellow-greens up to yellow and of the red-violets up to magenta, cyan divides into blue and into blue-green. The oranges and the reds have very little effect on hue, but they augment saturation. Simultaneous orange, which has no effect on cyan's hue, can be seen on page 73 (see also the summary on page 138).

Add 100% cyan.

Blue A at left below is made up of 30% green + 70% blue-violet. Blue-green B at right consists of 60% green + 40% blue-violet. The difference is of 30%, corresponding to the coefficient of simultaneous transformation of cyan on the opposite page.

Optical Creation of the Six Basic Colors.

To summarize, none of the six basic colors can be obtained through the optical mixing of equal parts of two primaries, as is the case with the subtractive system. The reason is that the brighter or more dynamic colors participate in the mixing with a diminished quantity — a phenomenon equally true of the complementaries.

Diagram of the quantitative ratios:

1. Blue-violet — magenta, + cyan
2. Yellow — red-orange, + green
3. Green — yellow, + cyan
4. Magenta — red-orange, + blue-violet
5. Red-orange — yellow, + magenta
6. Cyan — green, + blue-violet

A

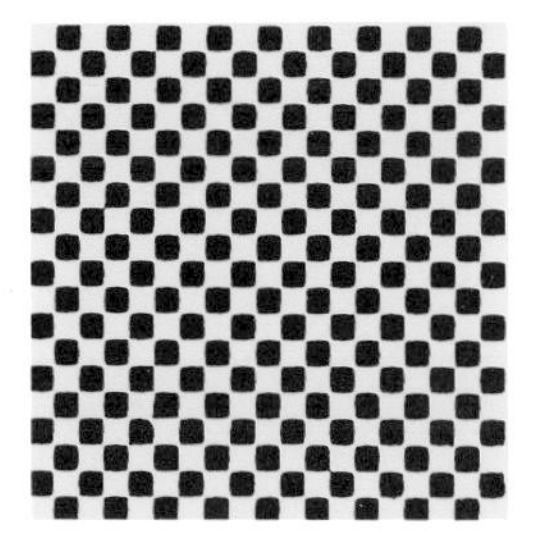

B

The Impressionists

Over a century ago, in 1874, the exhibition of the impressionist paintings of Monet, Renoir, Pissarro, Sisley, Cézanne, and Degas created a scandal. The critics of the time were full of sarcasm and disapproval. The term impressionist itself was originally an insult that the journalist Leroy lifted from the title of Monet's painting *Impression, Sunrise.* Breaking with the influence of Courbet and Corot, the broken tones of gray and green gave way to more luminous colors. In order to render reflections in water, the vibration of air and light, the colors were broken down into multiple facets of pure hues which are reconstituted only in the eye of the observer.

Chance effects of lighting, the transitory atmosphere of a landscape, dominated every compositional consideration. This total submission to appearances carried within it the danger of neglecting construction; it was for this reason that Degas and Cézanne associated themselves only briefly with the impressionist movement. Even Renoir confided to Vollard on one occasion: "No technique can be reduced to a universal formula."

The intuitive analysis of color would give way eventually to systematic experimentation. With Seurat and Signac, founders of the neo-impressionist group, artists became preoccupied as much with physiology and psychology as with vision and the optical problems of analyzing spectral luminosities. The discoveries of the physicist Helmholtz (1878), of O. N. Rood, as well as Chevreul's publication *The Law of Simultaneous Contrast,* which had appeared as early as 1839 (reprinted 1969),[7] were the basis on which they formulated their theories. Seurat's great composition, *Summer on the Grande Jatte,* created between 1884 and 1886, is characterized by a technique combining a multitude of points of color which he himself called "divisionism." The technique of replacing the mixing of pigments by optical mixing culminated in "pointillism."

Today the works of impressionism, divisionism, pointillism, and neo-impressionism are considered inestimable contributions to the evolution of contemporary art.

It is impossible to create saturated colors by optical combination. Only subtractive synthesis, by superimposing transparent inks, and additive synthesis, by adding spectral lights together, can create pure and luminous hues.

The preceding chapters have shown the limitations of the optical chromatic fan: the reds, the oranges, the greens, the blues and the violets resulting from an optical fusion are all desaturated either by white or by black.

For a color to be able to display all its brilliance, its surface should be large enough for the color to be perceived individually. The painting technique of juxtaposing colors by small brushstrokes permitted the impressionists not only to create bright or dark colors solely with pure and saturated hues, but also to create grays, as we will see in the following chapter.

The Complementaries

Definition of complementary:

1. *Additive synthesis:* two saturated complementary lights yield white when superimposed.

2. *Optical synthesis:* two saturated complementary colors (lights or pigment) are neutralized into an impression of gray when optically combined.

3. *Subtractive synthesis:* two saturated complementary transparent inks cancel each other out and become black when added together.[8]

A complementary can be called such only when it is capable of neutralizing another color into a white, a gray, or a black. In small doses (nonsaturated complementary) it desaturates without, like white or black, altering the hue.

The six basic colors of the subtractive system elicit six different optical complementaries. In order to neutralize, the eye effects a kind of correction in relation to the total physiological quantitative strength (see diagram 1b and disk a1, page 62).

Reduction of the magenta for the optical complementaries of the yellow and the cyan.

Reduction of the yellow for the optical complementary of the magenta.

Augmentation of the cyan for the optical complementaries of the blue-violet and the green.

Augmentation of the yellow for the complementary of the red-orange.

Overall, the optical complementaries elicit more blue-cyan and less red-magenta, as the relative percentages below indicate. Yellow remains more or less balanced in the middle, i.e., it constitutes about a third of the total.

The exact values given are, however, valid only for the inks used in this book. For example, the blue hue, optical complementary of the yellow, has in all cases the same appearance, but according to the primary inks used (specificity of cyan and magenta) the magenta part can vary between 50% and 70%, while it is only 50% with purer inks.

Optical complementaries

cyan	+ (100% yellow	+ 50% magenta)	= 40% black
yellow	+ (100% cyan	+ 65% magenta)	= 50% black
magenta	+ (100% cyan	+ 70% yellow)	= 60% black
red-orange	+ (100% cyan	+ 35% yellow)	= 60% black
blue-violet	+ (100% yellow	+ 40% cyan)	= 70% black
green	+ (100% magenta	+ 30% cyan)	= 70% black

Subtractive complementaries

cyan	+ (100% yellow	+ 100% magenta)	= 100% black
yellow	+ (100% cyan	+ 100% magenta)	= 100% black
magenta	+ (100% cyan	+ 100% yellow)	= 100% black
red-orange	+ 100% cyan		= 100% black
blue-violet	+ 100% yellow		= 100% black
green	+ 100% magenta		= 100% black

Along with the additive, optical, and subtractive complementaries, the existence of a fourth, the simultaneous or successive complementary, must also be acknowledged. Dependent on a translation, linked to a code (whether numeric or verbal), it is subject to errors of interpretation. While not objectively measurable, it is nevertheless possible to isolate it by systematic comparisons, staring at colors against neutral gray backgrounds of the same brightness, against white, black, or colored backgrounds, and then successively transpose the image onto a white screen. This phenomenon is not a mere physiological reflex but something that pertains just as much to memory and to the brain's analytical faculties.

This complementary is located midway between the subtractive and the optical complementaries. In certain borderline cases it resembles either the subtractive or the optical complementary (see the summary on pages 138 and 139).

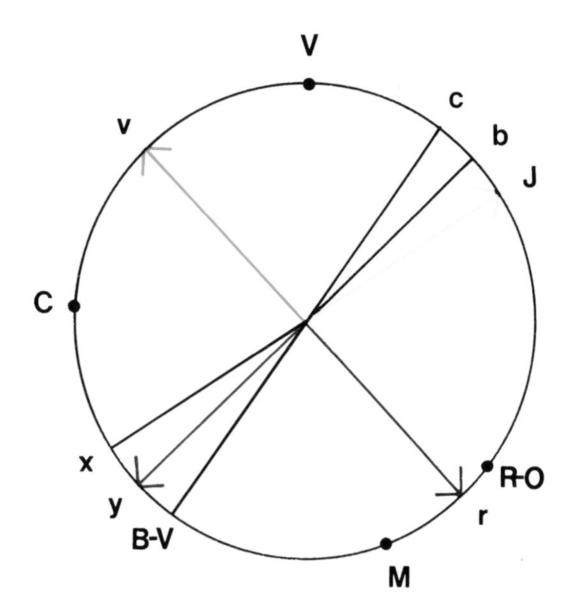

diagram 1

Diagrams 1, 1b, 1c, and 1d are conceived as follows: The circumference of a circle is divided into 48 equal parts, a transposition of the 24 pairs of optical complementaries from diagram 1a.

Diagram 1:

J = yellow: subtractive complementary of B-V (blue-violet). See page 17.

b = yellow-green: simultaneous complementary of B-V. See page 105.

c = yellow-green: optical complementary of B-V.

y = blue: simultaneous complementary of J, the yellow of page 113.

x = blue: optical complementary of J.

r = red: optical, simultaneous, and subtractive complementary of v.

v = cyanated green.

Diagram 1 at left shows the antagonist process in which the red/cyanated-green pair (r/v) is opposed at an angle of 90° to the blue (y) which is the simultaneous complementary of the yellow (J). In the (r/v) sector, the optical, simultaneous, and subtractive complementaries have a tendency to coincide (see page 85).

There can also be detected an oscillating movement of the yellow's complementaries grouped around y toward the optical complementary at x on the one hand, and toward the subtractive complementary at B-V on the other.

The contrasts between the three kinds of complementary are greater for all the blue hues opposite the hues between red-orange and green. They are relatively unimportant for the reds around r opposite the blue-greens around v.

Why is there more blue and green than red or orange in the optical complementaries as a whole (see the diagrams and disks on pages 62 and 63)? The following hypothesis appears the most probable: the eye mirrors not only physiological interior but the exterior world with which the body is interdependent. In the long history of animal evolution, the adaptation of the eye took place originally in relation to the cycles of light and dark in the alternation of day and night; and subsequently in relation to a polychromatic environment in which the division of colors is determined quantitatively: the blues of the sea (first habitat of man's ancestors) and of the sky predominated. Seen from space, Earth is a blue planet. Afterwards come the greens (second habitat of our ancestors). Rocks, soil, sand display beiges, ochres, browns — all of them desaturated hues. In nature, blues and greens unfold in all their brilliance; conversely red, orange, and yellow are seldom found in saturated form among fruits, flowers, certain fish, a few birds, and within the human body, in the blood. Likewise the human skin contrasts with the dominant blue-green, ranging from "white" to "black" by way of beiges, pinks, light browns, and deep browns.

It is true that the sky is not always blue. It can equally as well be gray; rare are the euphoric moments when sunsets deck the sky in sumptuous oranges, reds, and purples.

We will see on page 63 that red-orange, green, and blue-violet must be combined in precise proportions to achieve, on a rotating disk, a neutral gray of about 70% black. Comparing these proportions with diagram 1c, the resemblance is striking. The analogy between relations of quantity and global perception, that is to say the positioning of the totality of the complementaries and the neutralizing of the three additive primaries, is no fortuitous coincidence.

The location of the cones sensitive to red-orange and green toward the center of the retina, and of those sensitive to blue toward the periphery, mixed with rods, is supposed by Walraven to be in the following average proportions: red/green/blue: 40/20/1, and, according to Vos: 32/16/1. These quantities put red in the majority and blue in the minority; in other words, the reverse of the proportions found in nature and discussed above. It is logical that the visual apparatus should be more developed for dealing with the colors which are rarer in visible reality, since they have to be distinguished from a background in which blue and green prevail, and were therefore of great importance for human survival.

These physiological arrangements may explain to some extent the extra sensitivity of the eye to warm and saturated hues, which draw the attention even on small surfaces and become unbearably aggressive in large quantities (for example, a room painted entirely in red-orange). This emotional reaction is thus not a psychological factor attributable to the influence of a culture or an era; it appears to spring rather from our hereditary make-up.

diagram 1a

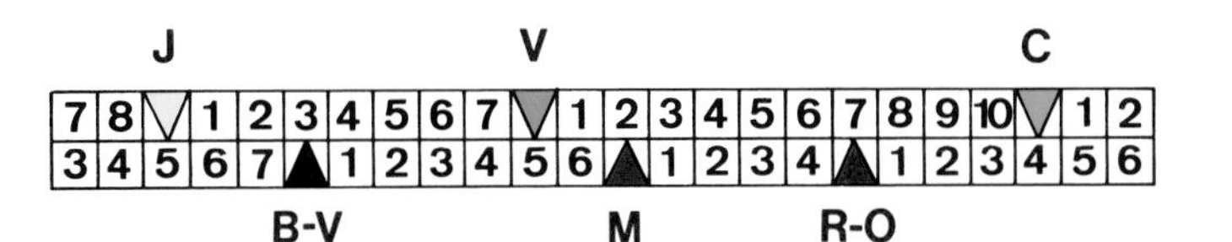

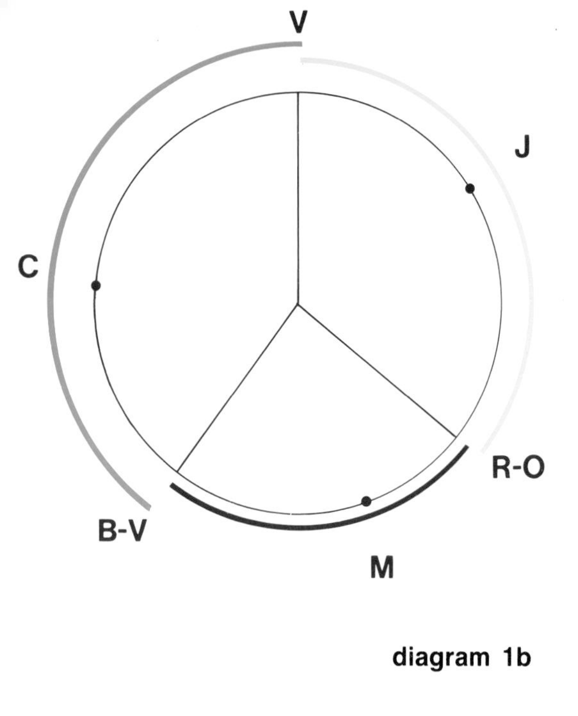

diagram 1b

a1

b1

The respective quantities of magenta, yellow, and cyan for the totality of the optical complementaries:

Diagram 1b:

- 24% of: 100% magenta
- 36% of: 100% yellow
- 40% of: 100% cyan

Saturated magenta: all the colors ranging from blue-violet to red-orange

Saturated yellow: all the colors ranging from red-orange to green

Saturated cyan: all the colors ranging from green to blue-violet

Compare these proportions with those of disk a1.

Optical synthesis of the three subtractive primaries.

Magenta, cyan, and yellow optically combined in the proportions opposite neutralize into gray, whether by turning disk a1 or by juxtaposing small dots in these precise quantities.

Disk b1 = disk a1 in rotation, with a gray of about 40% black.

Disk a1:

- 20% of: 100% magenta
- 27% of: 100% yellow
- 53% of: 100% cyan

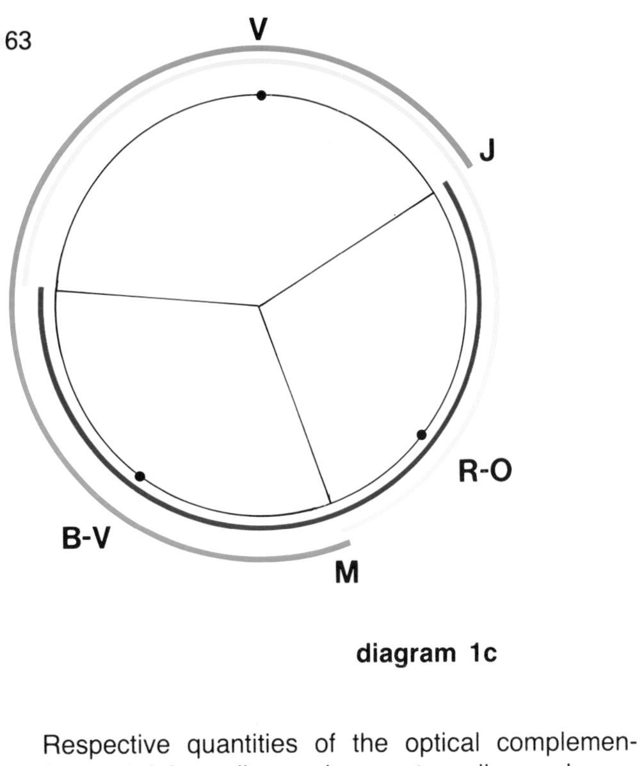

diagram 1c

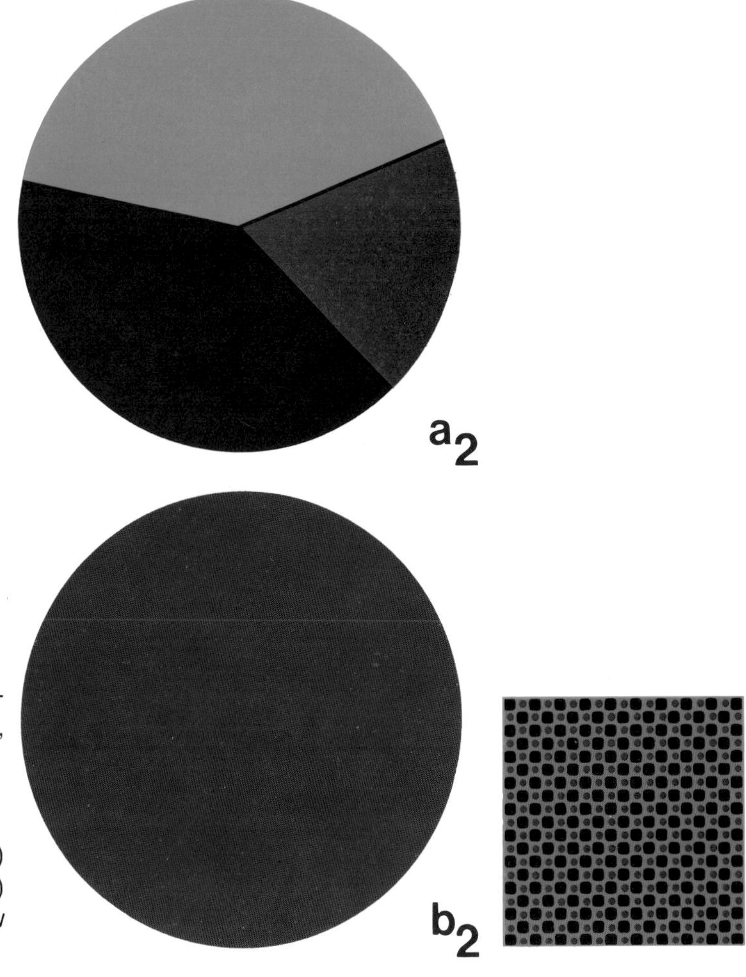

Respective quantities of the optical complementaries containing yellow and magenta, yellow and cyan, cyan and magenta:

Diagram 1c:

28% of: reds and oranges (magenta to yellow)
32% of: blues and red-violets (magenta to cyan)
40% of: yellow-greens and blue-greens (yellow to cyan)

Compare these quantities with those of disk a2.

These precise quantitative ratios (disks a1 and a2) obviously apply only to the six primaries used in this book. Nevertheless it can be affirmed that gray can never be obtained from equal parts of the three respective primaries, but that cyan dominates considerably, followed by yellow, with magenta in the minority on disk a1.

Disk a2 is divided into red-orange, blue-violet, and green in the proportions below. Optical mixing by rotation (see disk b2) brings out a deep gray (about 70% to 80% black).

Disk a2:

19% of: 100% red-orange
40% of: 100% blue-violet
41% of: 100% green

According to the variations in reproduction, it will ultimately be necessary to readjust one of the three components by 1% to 2% more or less in order to obtain a perfect neutrality.

Twenty-Four Pairs of Optical Complementaries

To optically neutralize two colors, attention must be paid to two variables: on the one hand the hue (A), solely responsible for the choice of complementaries; on the other hand, the saturation, brightness, or the dynamics which establish the respective quantity of the two complementaries (B).

A. The hue depends on the percentage, whether more of less, of a second primary added to a 100% saturated primary. The continuous traits of yellow, cyan, or magenta indicate 100% saturation; the numbers alongside give the percentages of the second primary component responsible for the hue.

The number of gradations between two colors seems arbitrary — it is not, however, infinite. For example, in printing, it is practically impossible to go below a graduation of 5%-10%-15%-20%, etc., up to 100%. The danger that the differences will become imperceptible below 5% exists even in the scale of yellows, yellow-oranges, and yellow-greens, where the degree of differentiation is nonetheless highest. The reds on page 66, spaced at gradations of 20% (100% magenta + 10%, 30%, 50%, 70% and 100% yellow), already give an impression of continuity, compared to the blue-greens which are spaced only at 10%. Cyan respects yellow's differences of hue in the blue-greens better than magenta does for the reds.

The length of a scale depends on the position of the complementaries, as indicated in diagrams 1a, b, c and d. The greens and the blue-greens in particular predominate over the blues, red-violets, and oranges.

20 10 10 25 40

J

B-V

35 50 65 75 85

They extend over the borders of the oranges on one hand and of the red-violets on the other. The reds are in a minority.

The darkest chromatic value, blue-violet, the threshold where magenta and cyan are both at their maximum, is opposite a sector in which spectral luminous efficiency is at its most sensitive: yellow-green.

There is a different spacing disposal of the complementaries compared to diagram 2, which opposes the six thresholds of chromatic sensitivity to each other, and which holds good for the subtractive complementaries.

diagram 2

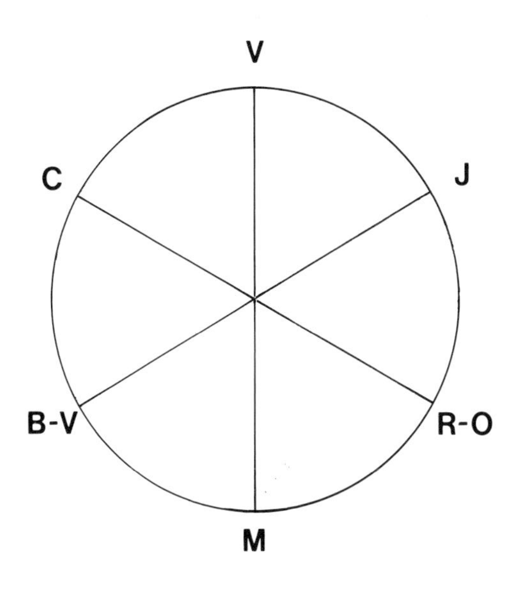

50 60 70 80 80

80 65 50 40 30 15

B. Quantitative ratios for making a neutral gray by optical synthesis:

1. The brightest saturated complementary is in the minority.

Examples:
24% white : 76% black; 26% yellow : 74% blue; 30% yellow-green : 70% blue-violet.

2. When two saturated complementaries are of the same brightness, the percentage of the warmer color is reduced.

Examples:
35% magenta : 65% blue-green; 41% red-orange : 59% blue-green

3. The only complementary pair which can achieve neutralization in equal mixture consists of two colors of the same brightness and exactly midway between warm and cold: 50% green : 50% red-violet.

These quantitative ratios of optical complementaries go hand in hand with the opposite dynamizing qualities of the colors:

Bright colors approach the observer. White and yellow, for instance, radiate outward. The warm, active colors, the reds and oranges, also move outward, culminating in the aggressiveness of red-orange.

Dark colors and cold colors both recede. Blue in particular creates an atmosphere of calm and detachment, a psychological factor which can be explained by means of these objective characteristics of the expressiveness of colors.

70 60 55 50 40 35

10 30 50 70

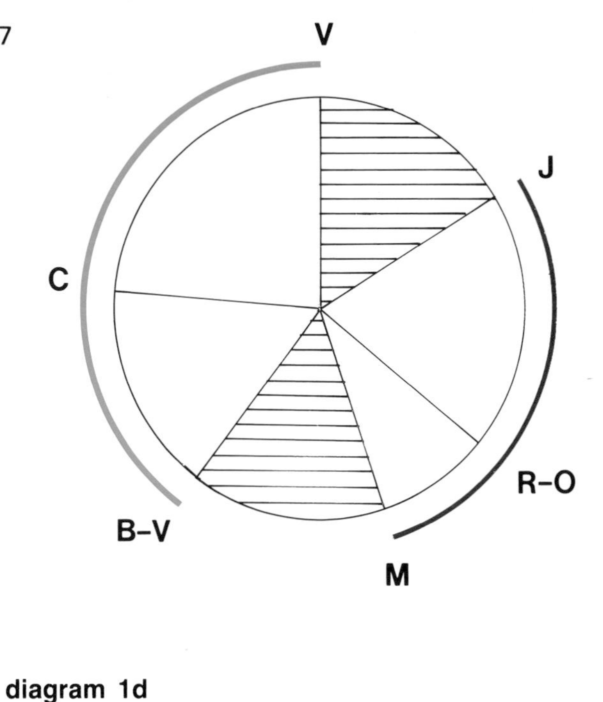

Diagram 1d

Diagram 1d shows the quantitative ratio of the cold optical complementaries as opposed to the warm optical complementaries, as well as the position of the optical complementaries which are relatively either warmer or colder.

Between V (green) and B-V (blue-violet) by way of C (cyan) are located the greens, blue-greens, and blues, which are in essence psychologically cold colors.

The colors between J (yellow), R-O (red-orange) and M (magenta) are classed as warm. The red-violets between B-V and M, as well as the yellow-greens between J and V, are colors which can be either warm or cold depending on their context.

diagram 1d

30 20 10 10 20

C

80 70 60 50 40 30

Nonsaturated Optical Complementaries

We see above three scales of gray, ranging from bright to dark. They are made up of three complementary pairs.

The grays on the opposite page are not perfectly neutral, because the quantitative ratios in use are only approximations, due to the difficulties of reproduction.

What happens when a complementary is brightened or darkened? The scales on this page indicate clearly the respective quantities of yellow, cyan, and magenta. These quantities decrease proportionally to desaturation toward either white or black, starting from a maximum located opposite the purest saturated complementary.

There are inks which are purer and consequently of higher saturation. These are colors made from pigments rather than composed of two primaries like the ones in use here. Alongside such vivid hues, some of the colors in this book — notably the blue-violet and the neighboring blues and red-violets — appear darker and duller.

The "cleanest" hues of the scale on pages 64 through 67 are the blues cyanated with cyan containing turquoise, as well as the reds.

By way of example, the table below shows the degree of blackening of the six 100% saturated colors printed on these two pages, in comparison with the corresponding hues as presented in a color lexicon:[6]

Yellow	=	2 A 8 (0% black)
Blue	=	20 D 8 (20 A 8 + 40% black)
Cyan	=	23 A 8 (0% black)
Orange	=	5 B 8 (5 A 8 + 10% black)
Magenta	=	13 A 8 (0% black)
Green	=	25 C 8 (25 A 8 + 30% black)

Example:

The green of 100% (relatively pure and saturated as we have stated above), the most intense hue of scale c, is opposite the greatest quantity of magenta, 32%-35%. A bright green of 30% needs only 25% saturated magenta to become a bright gray, and only 15% saturated magenta (quantity of the magenta dots) against a darker background (saturated green + 30% black) to become a dark gray.

The average and relative ratios indicated on the disk a on the following pages for saturated pairs of optical complementaries can vary according to the purity and luminosity of the hues which are combined. In order to compensate for a higher degree of limpidity, the quantity of the optical complementary should be augmented.

A

100 95 90 80 70 6

Cyan and Orange

How can we find the optical complementary of cyan?

The optical complementary of cyan is an orange, but how much magenta has to be added to the yellow? A scale of oranges — 100% yellow + 10%, 20%, 30%, 40%, 50%, 60%, 70% magenta — will allow us to make a selection.

If, on our disk, we rotate cyan with an orange which is more or less red, it will be necessary to eliminate all tendencies either toward red-violet or toward green. According to the characteristics of the primary cyan ink, the complementary orange can vary between 30% and 50% magenta added to the saturated yellow. For the cyan used in this book, the orange must have 50% magenta.

50 40 30 20 10 0

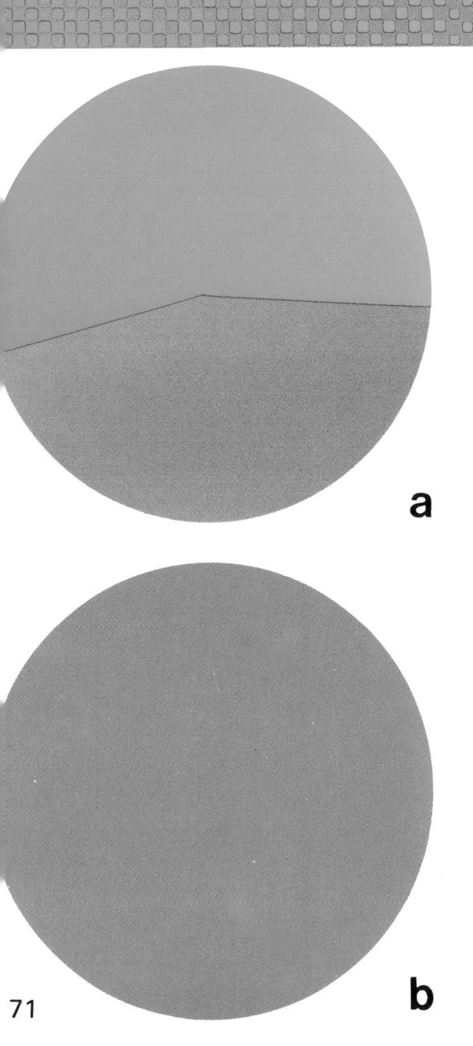

55 parts cyan + 45 parts orange = gray (40% black).

The percentage of orange can vary from 35 to 47 depending on the saturation of the cyan.

The optical complementary poles cyan and orange are at opposite ends of the scale above. The orange is desaturated in five stages moving toward intermediate gray. At each stage it becomes slightly deeper.

Cyan on the other hand becomes increasingly grayer in only six stages: 95%-90%-80%-70%-60%-50% to 53% (see the place marked by the arrow).

Disk a indicates the same ratio. When rotated, the disk shows a medium gray represented by disk b.

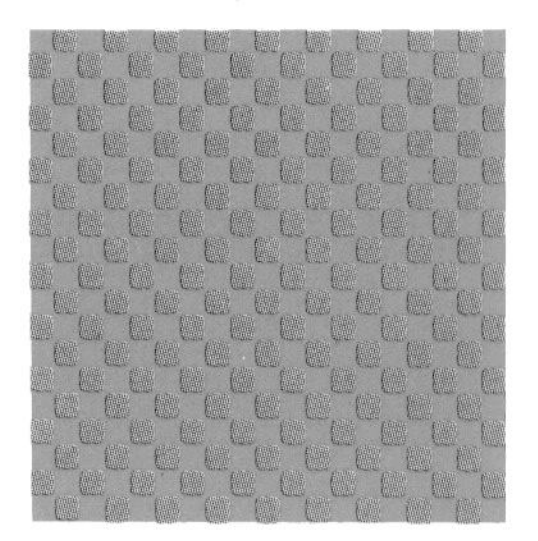

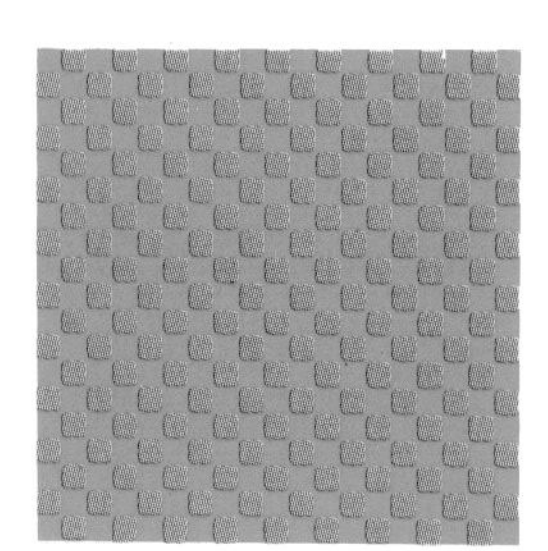

Simultaneous Effects: Cyan and Orange

In order not to distract your attention when comparing the two small squares above with those on the opposite page, it is best to cover the two circles with a sheet of paper. From far enough away, these two small gray squares resemble, in terms of saturation, the two squares in the center of the cyan background and the orange background. They are darker because of the simultaneous effects of the surrounding white, much brighter than the backgrounds on the opposite page.

An orange and a blue both appear gray.

Maximum desaturation (opposite page, above): the cyan background brings out a simultaneous orange (lower disk on the page at left) which results in the weakening of the blue elements in the small blue square at left (cyan + orange = gray). The blue (semicircle below, right), which is the simultaneous complementary of the orange background, on the contrary fortifies the blue in the small orange square at right (orange + blue = gray).

The simultaneous complementary of cyan (100% yellow + 75% magenta) is redder than its optical complementary.

The simultaneous complementary of orange is a blue containing magenta (100% cyan + 25% magenta).

As proof, the successive image reveals a perfect inversion of the two pairs opposite.

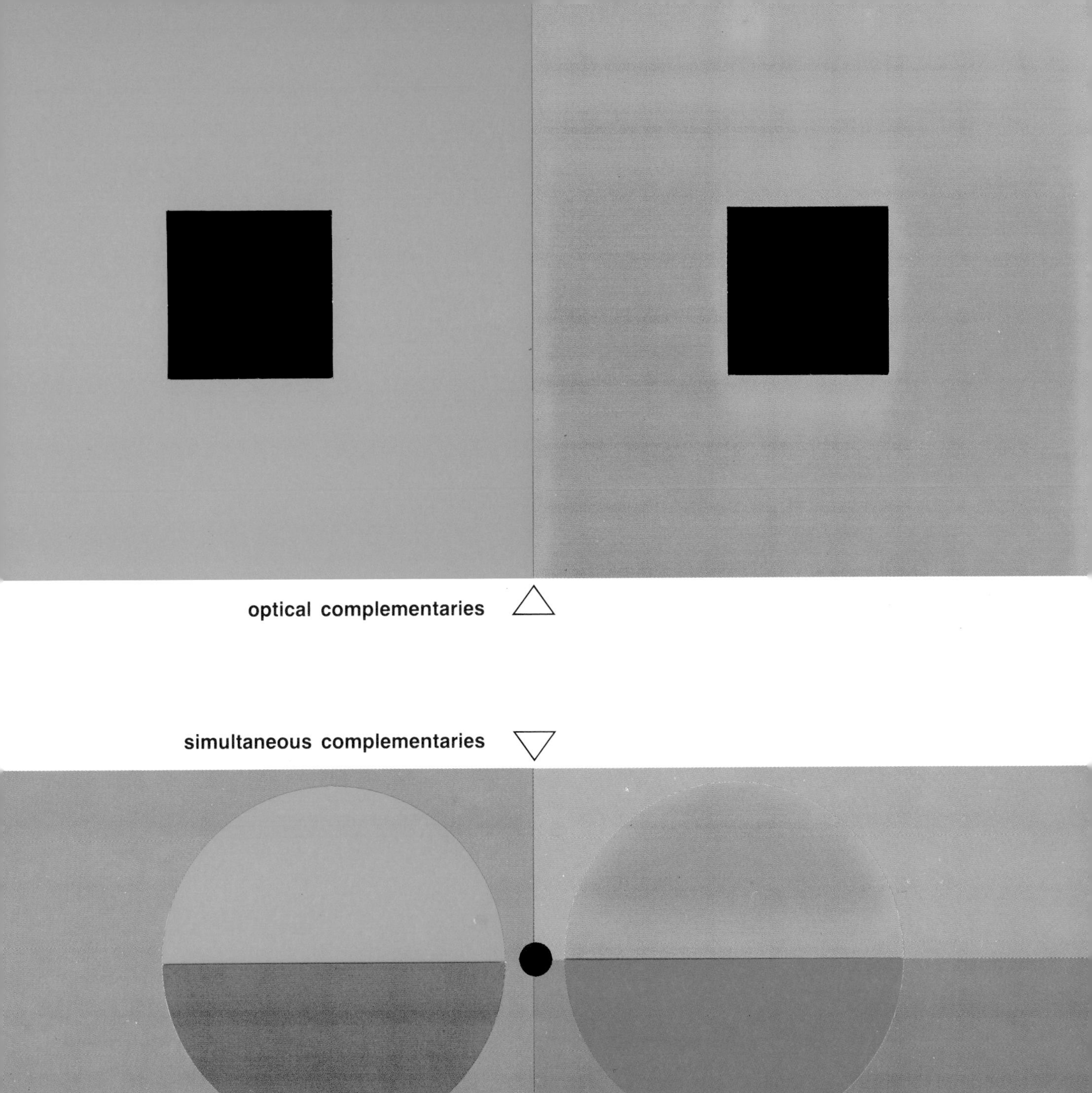
optical complementaries
simultaneous complementaries

Cyan + orange = optical gray.

The dots are divided into 40% orange and 60% cyan. With a background consisting of 40% black they become indistinguishable.

Maximum reinforcement of saturation.

On the opposite page the surrounding cyan exaggerates the luminosity of the central orange, while the orange background reinforces the brilliance of the blue. There is a reversal of the simultaneous effects on the preceding page.

A gray appears blue or orange.

Cleavage of gray into its two constituents: a gray with a blueish tendency in the center of the orange, with an orange tendency in the center of the cyan. The coefficient of simultaneous influence of 30% corresponds to the difference between the two squares on the opposite page. (Compare with the positions on the scale on pages 70 and 71).

Page 77:

70 parts blue + 30 parts orange = the blue square at left.

30 parts blue + 70 parts orange = the orange square at right.

A

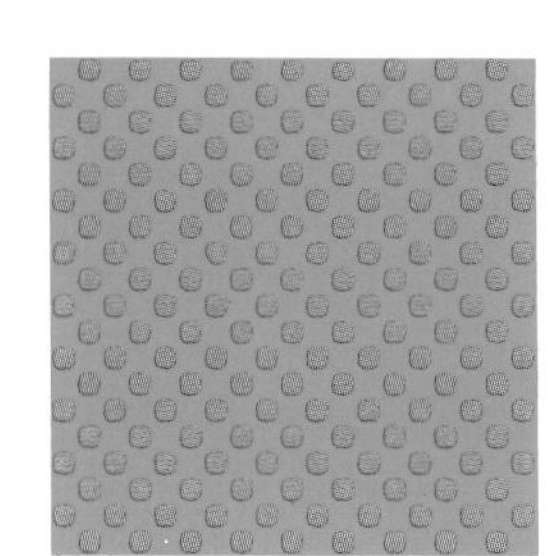

B

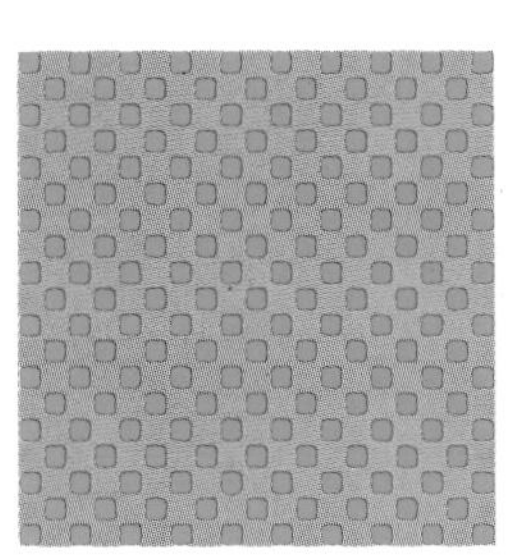

A

100 90 80 70 60 50

Turquoise and Red-Orange

The dynamic factor of red-orange.

The optical pair red-orange/turquoise proves that it is not only luminosity and purity which are decisive in the quantitative ratios of optical synthesis, and that there is a dynamic factor (warm color) which dominates over the characteristic of brightness. Red-orange has a 10% degree of blackening by comparison with the purest and most luminous turquoise, and yet it is red-orange which must be in the minority in order to achieve neutralization. The term "warm color" may lead to confusion, because in physics it is the short-wavelength, high-frequency colors which are called "warm".

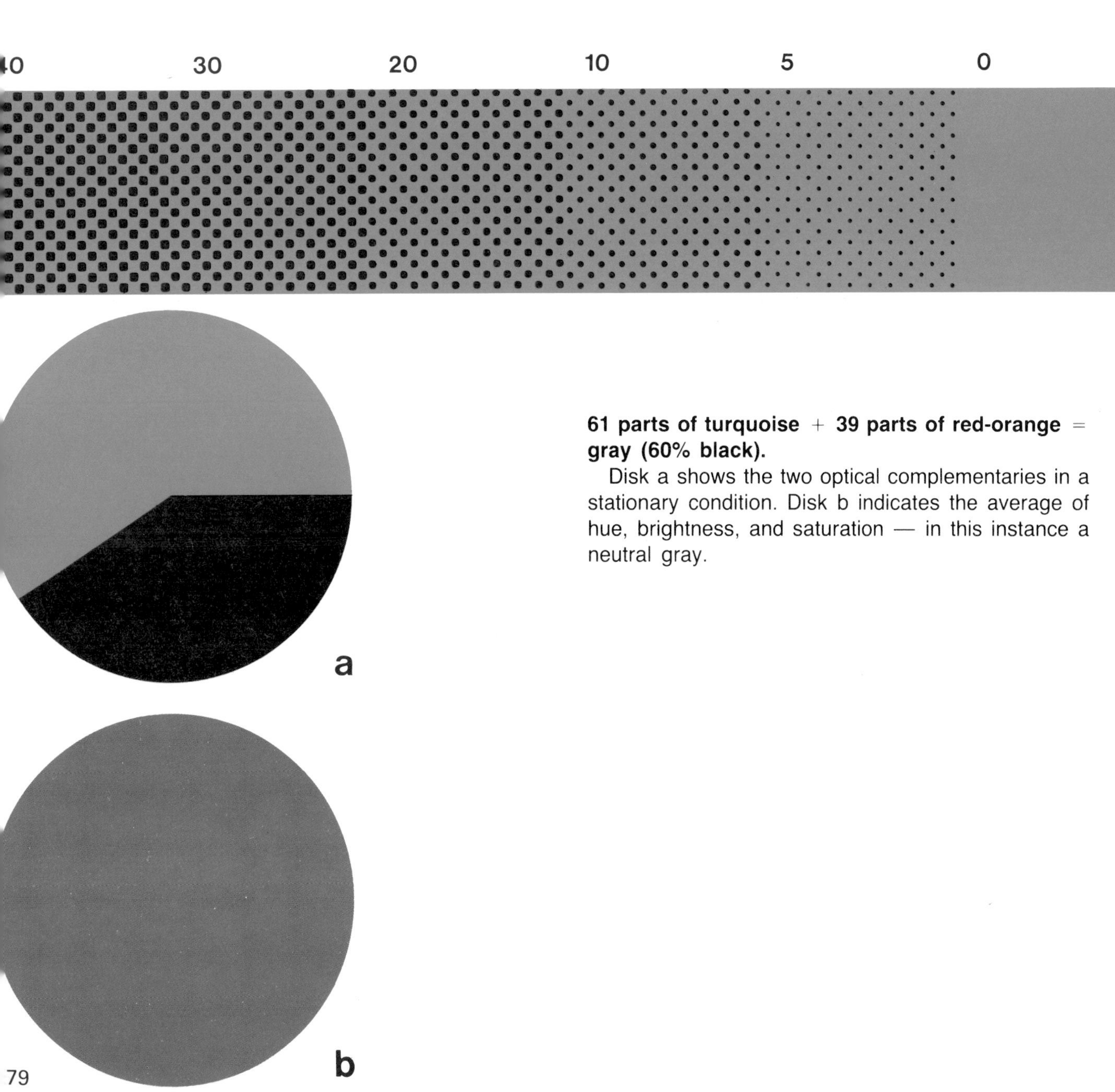

61 parts of turquoise + 39 parts of red-orange = gray (60% black).

Disk a shows the two optical complementaries in a stationary condition. Disk b indicates the average of hue, brightness, and saturation — in this instance a neutral gray.

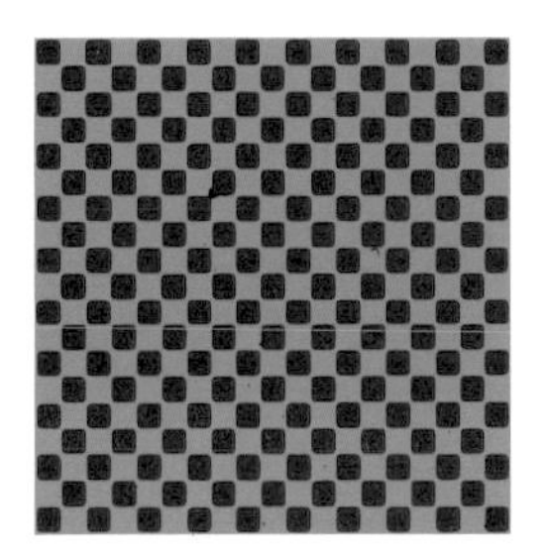

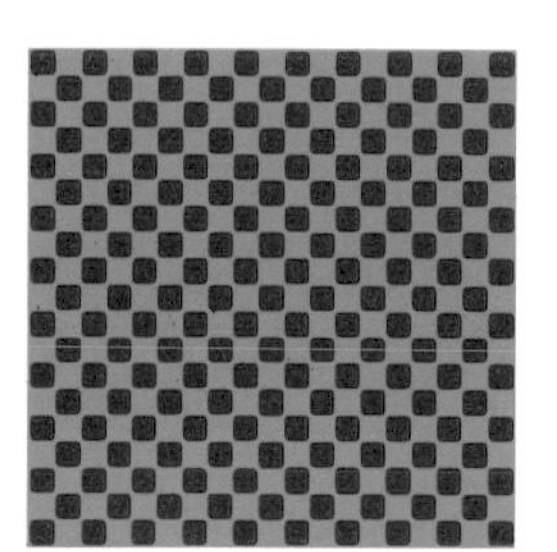

Simultaneous Effects:

Blue-Green and Red-Orange

From far enough away the two squares on the opposite page resemble the two squares surrounded by red-orange on the left and by blue-green on the right.

A red and a blue-green both appear gray.
Maximum desaturation.

A brilliant color of the same hue has the maximum reductive effect on the saturation of the color which it surrounds. Thus the brilliant blue-green background activates the simultaneous red-orange which, when it combines optically with the non-saturated blue-green in the center, makes it appear even grayer.

On the other hand, the surrounding red-orange floods the central square with a simultaneous blue-green and at the same time pushes the desaturated red toward gray.

Blue-green and red-orange = optical gray.
Opposite page:

The dots are divided into 60% blue-green + 40% red-orange. With a background of 60% black, they become indistinguishable. The turquoise dots lack 5% yellow, in order to achieve a perfect gray when mixed with this red-orange.

Maximum reinforcement of saturation.

The blue-green background relates the simultaneous effects of its complementary to the central red-orange, which glows with more luminosity than the color surrounded by neutral white on page 85.

In the same fashion the red-orange background stimulates the purity of the elements corresponding to its simultaneous complementary: blue-green.

By simultaneity a gray appears blue-green or red-orange.

Cleavage of the gray into its two complementary components: the gray surrounded by red-orange appears blue-green; the gray surrounded by blue-green appears red-orange. The coefficient of simultaneous influence of 30% corresponds to the difference between the two colors on the opposite page.

40% blue-green + 60% red-orange = red (A) on the opposite page.

70% blue-green + 30% red-orange = blue-green (B) on the opposite page.

The red and the blue-green on the facing page have a privileged position. This pair is located in an area in which all three types of complementaries have a tendency to coincide.

1. Subtractive complementary:

100% magenta	+	50% yellow		
+ 100% cyan	+	50% yellow		
100% blue-violet	+	100% yellow	=	black

2. Optical complementary: 40 parts red optically mixed with 60 parts blue-green = gray (about 60% black). The red component can vary between 38% and 41% depending on the degree of saturation of the green.

A

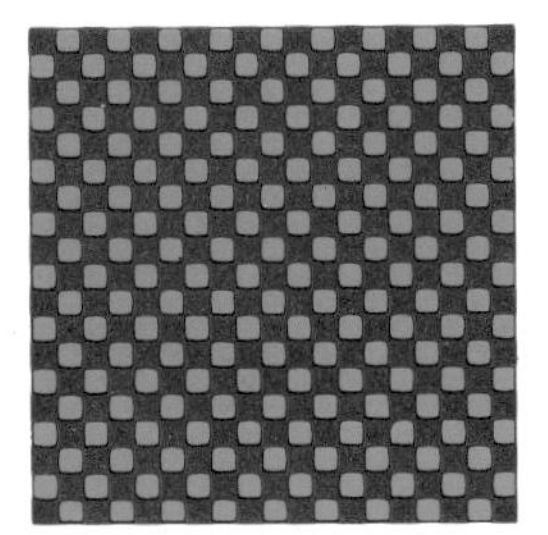

B

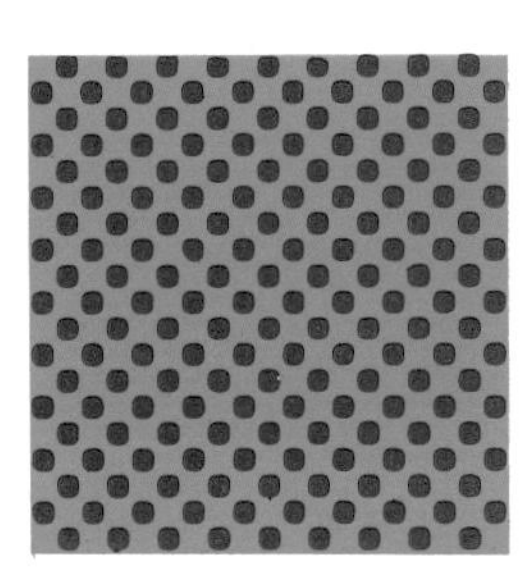

3. Simultaneous or successive complementary: you can see the proof for yourself by first concealing all the other colors with white sheets of paper, with the exception of the red-blue-green pair opposite against a gray background.

Stare at the central point for about 30 seconds, then project the successive image onto a white screen above you. If you perceive an exactly reversed pair, successive complementarity is proved.

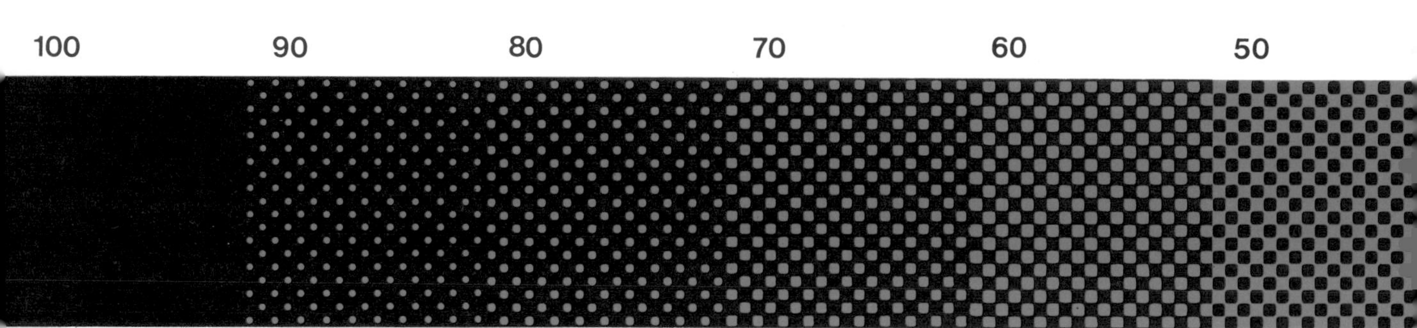

Green and Magenta

The optical complementary of magenta.

Is a yellow-green or a blue-green more suitable for turning magenta into a gray? A green scale (100% cyan + 40%-50%-60%-70%-80%-90% yellow) combined optically with magenta will provide the answer. We are on the right track if there is not too much orange on one hand or too much red-violet on the other. The ideal solution with the primaries used in this book turns out to be around 100% cyan + 70% yellow.

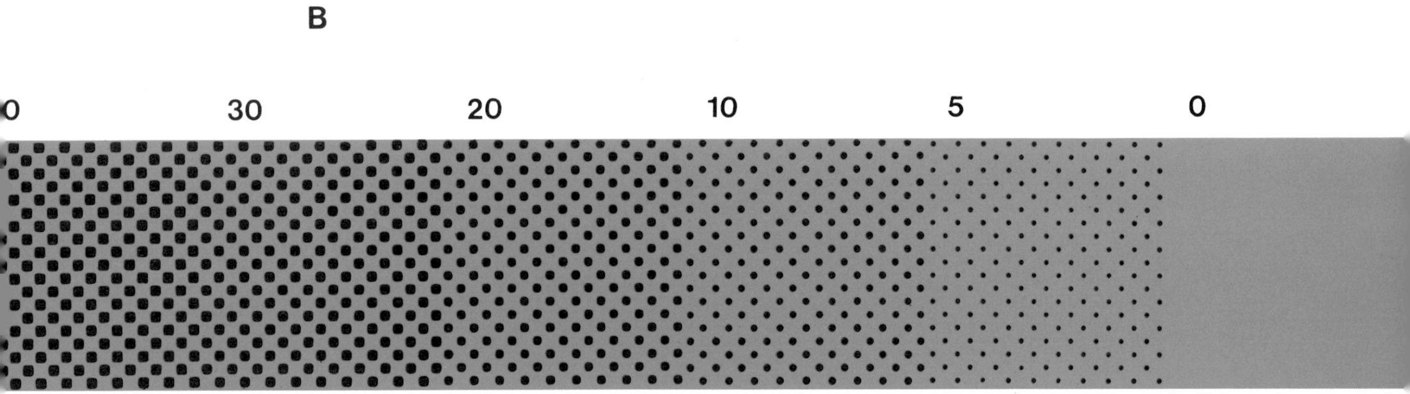

B

0 30 20 10 5 0

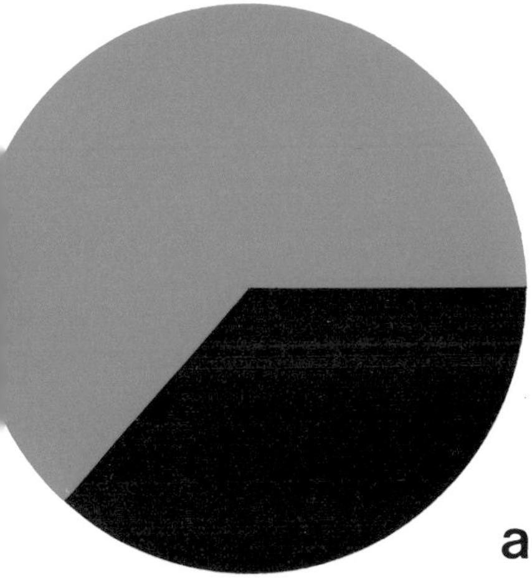

a

b

65 parts blue-green + 35 parts magenta = gray (60% black).

The brightness of saturated magenta is very close to that of its optical complementary, green. It is a warm color; as with red-orange it requires only, following a saturation of the green, between 32% and 40% magenta (see disk a) to become, through rotation (see disk b), a neutral gray.

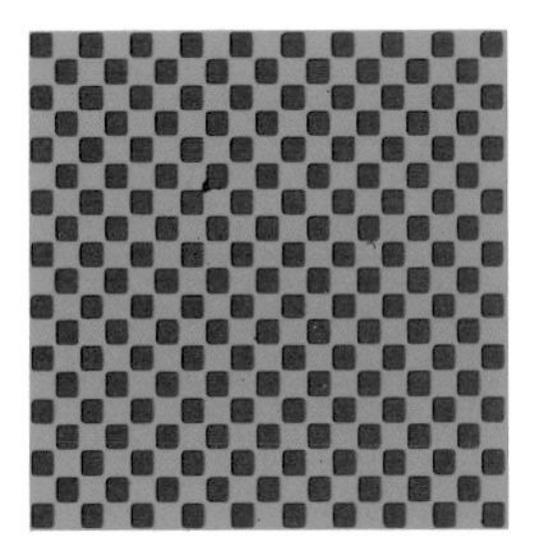

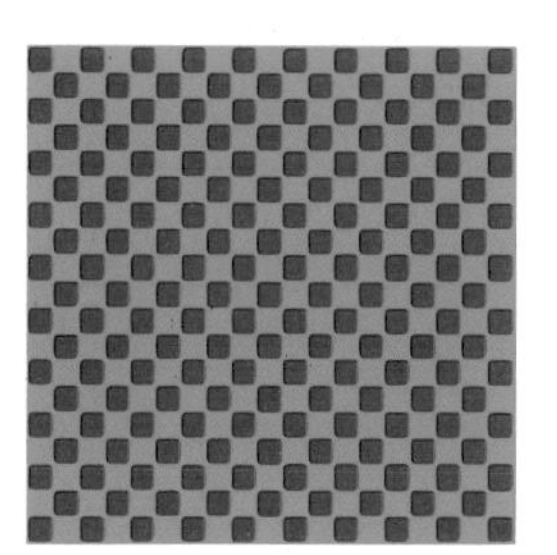

Simultaneous Effects: Green and Magenta

A red and a green appear gray.

The two gray squares on the opposite page resemble in color and saturation the two squares in the center of the magenta background and of the blue-green background above.

Maximum desaturation.

The red at left appears gray because the magenta background weakens with its simultaneous complementary magenta elements, while emphasizing the green elements.

The green background has the maximum impact in terms of desaturation on all the desaturated greens located between gray and 100% green.

Green + Magenta = Optical Gray.
Opposite page:
The dots are divided into 65% blue-green and 35% magenta. With a background consisting of 60% black they become indistinguishable.

Contrast of hue and increased saturation.
The surrounding magenta sustains the luminosity of the central green, while the green background sustains the brilliance of the magenta. The simultaneous effects on page 89 are reversed.

A gray appears red or green by simultaneity.

The gray tends towards magenta in the center of the green and toward green in the center of the magenta.

40% green + 60% magenta = square A on the opposite page.

70% green + 30% magenta = square B on the opposite page.

The simultaneous coefficient of 30% corresponds to the difference between the two colors red and green on the opposite page.

A

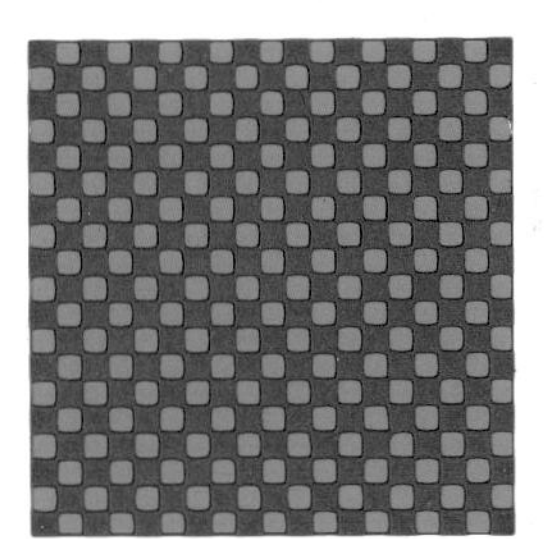

B

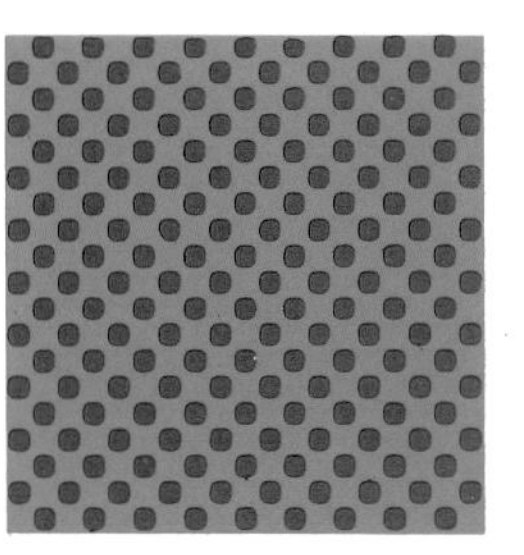

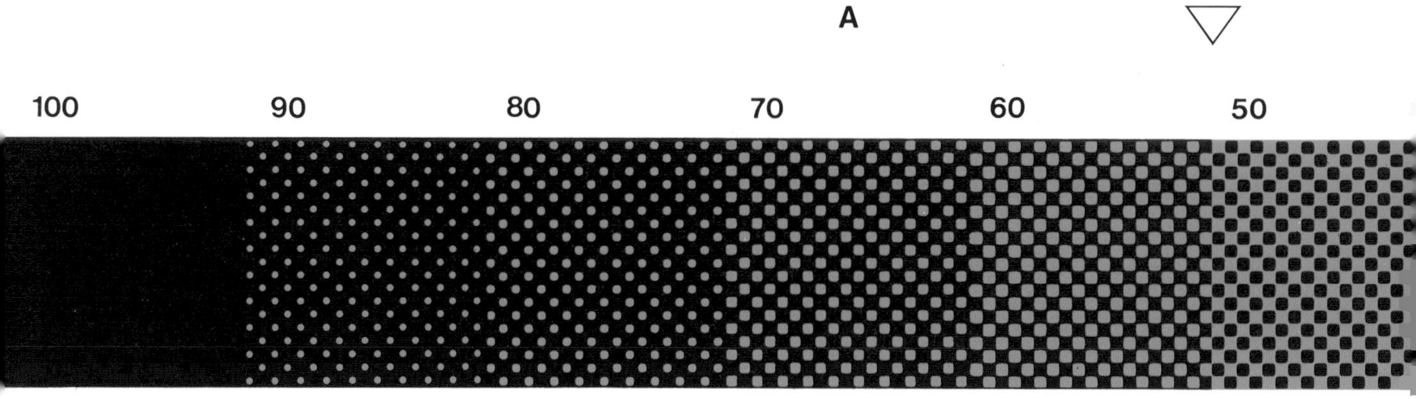

A

100 90 80 70 60 50

Green and Red-Violet

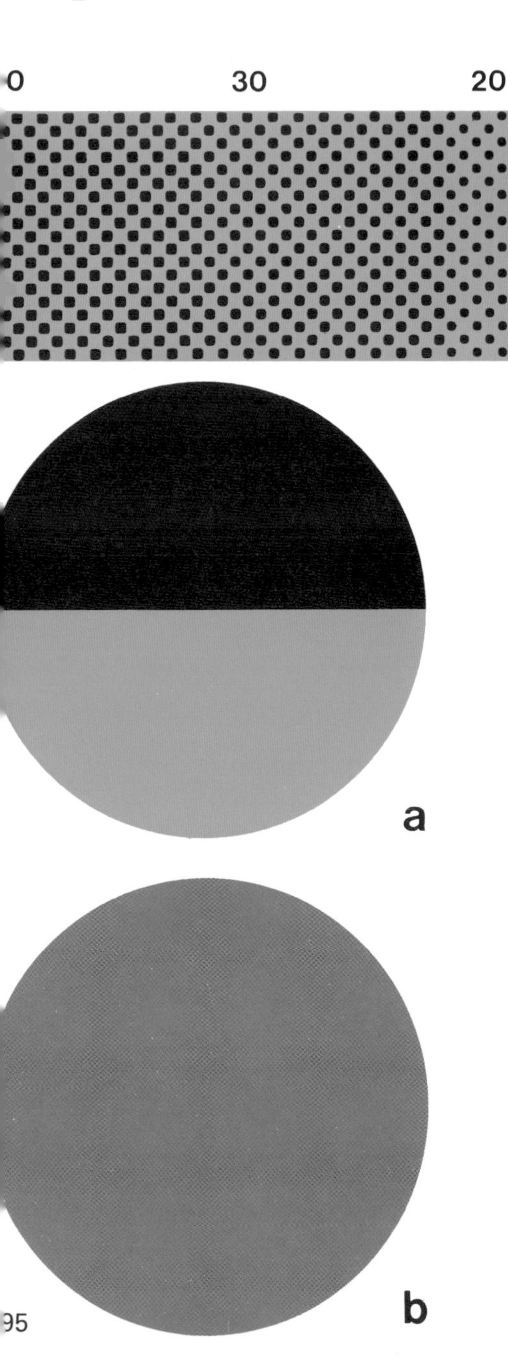

50 parts red-violet + 50 parts green = gray (65% black).

This is the only scale in which the neutral point is exactly midway between the two optical complementaries. It is a rather unusual pair: both being of the same brightness and either warm or cold depending on context, the red-violet (100% magenta + 40% cyan) and the green (100% yellow + 80% cyan) are in perfect equilibrium.

The gray (70% black) on disk b shows their optical combination by the rotation of disk a. The red-violet of the scale above, as well as that on page 99, corresponds more to the desired hue than the red-violet on disk a, which is a little too blue.

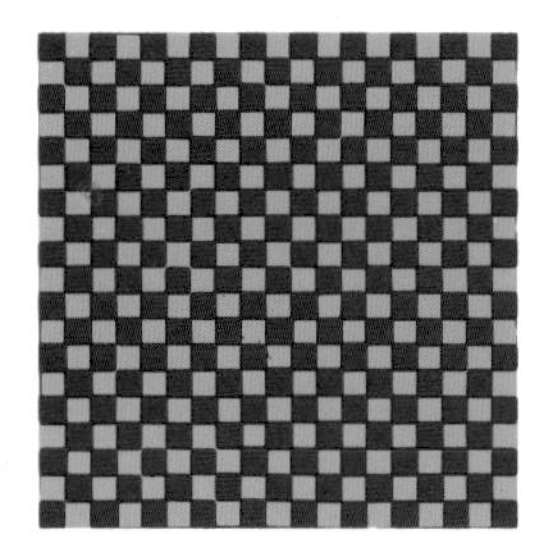

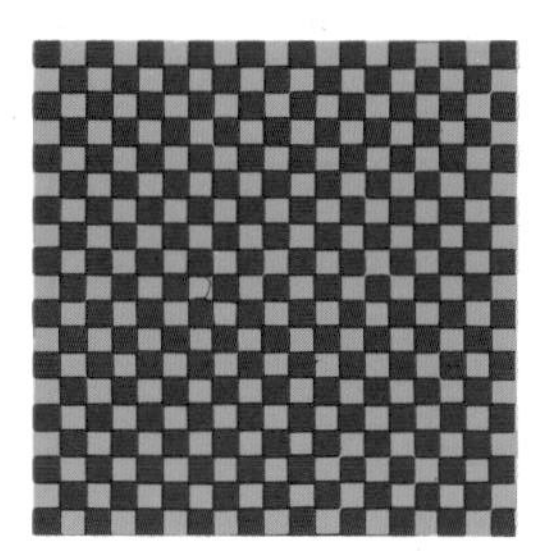

Simultaneous Effects: Green and Red-Violet

A red-violet and a green appear gray.

From far enough away, the two small squares on the opposite page resemble in saturation and hue the two squares surrounded by red-violet and by green respectively. As far as brightness is concerned, they are, of course, darker because seen against a brighter background.

Maximum desaturation of the red-violet square by the pure red-violet background; the green square is desaturated by the pure green background.

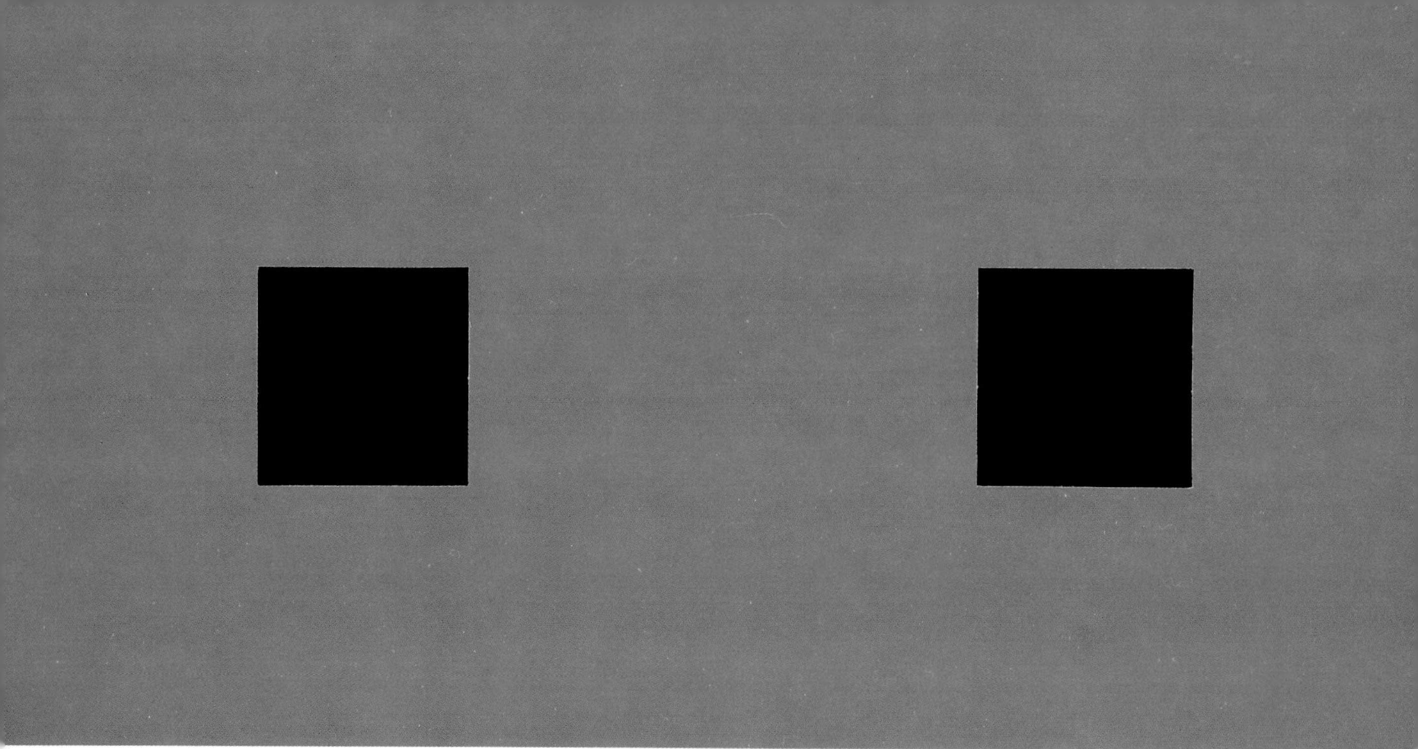

Green + Red-violet = Optical Gray.
Page 98:
50% red-violet dots combine optically with 50% green dots. They become indistinguishable against a gray background (80% black).

Maximum reinforcement of saturation.
Here the simultaneous effects on page 97 are reversed. The red-violet complementary accentuates the purity of the central green — while the green background is complementary to the red-violet square in the middle.

Cleavage of gray into its two constituents.

The gray surrounded by red-violet appears green. The gray surrounded by green appears red-violet. The coefficient of simultaneous influence of 30% corresponds to the difference between the two colors on the opposite page. (See the gradations of the scale on pages 94–95.)

A

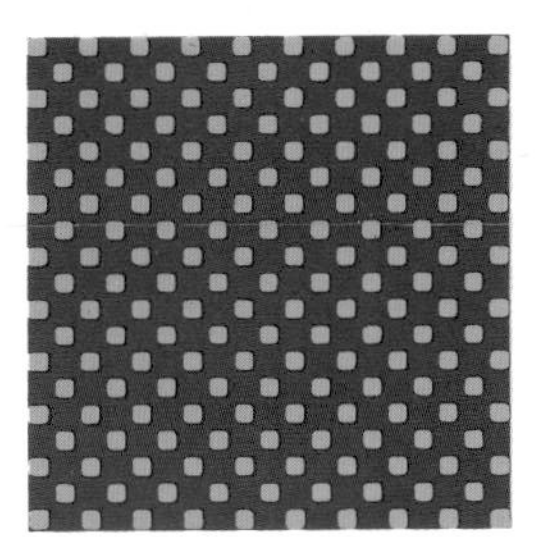

B

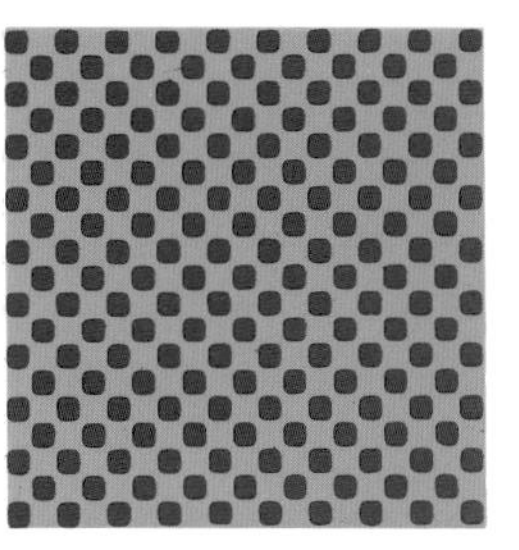

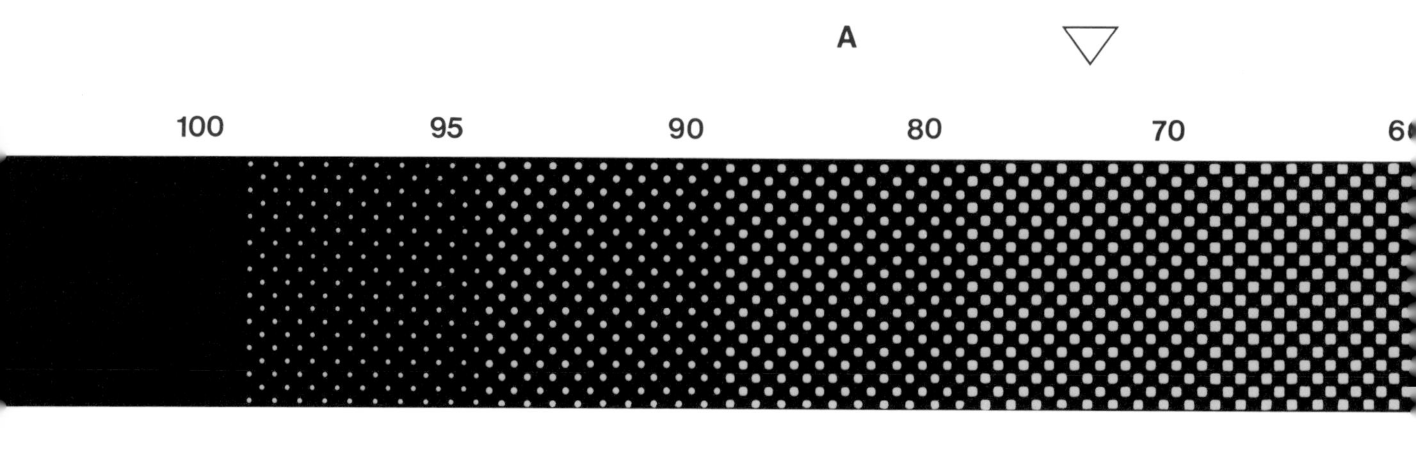

Blue-Violet and Yellow-Green

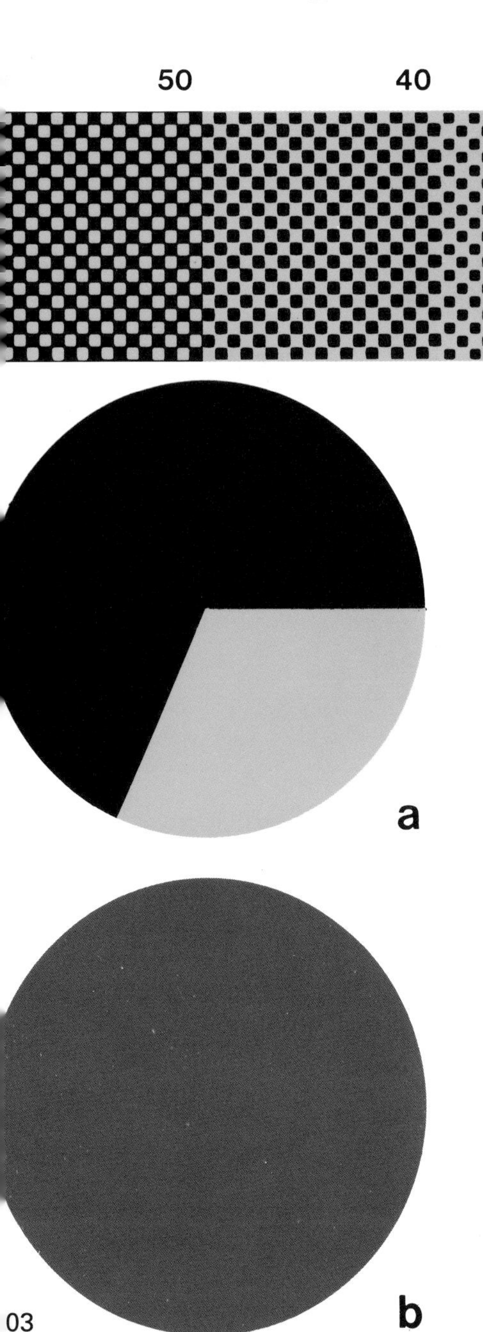

70 parts blue-violet + 30 parts yellow-green = gray (70% black).

There is a very marked contrast in brightness between yellow-green and blue-violet, the deepest of all the saturated hues. Consequently a section of only 30% yellow-green combined optically with 70% blue-violet results in the deep gray on disk b.

The scale above recalls the fact that blue-violet (100% cyan + 100% magenta) modulates yellow-green in the same way as black.

The two complementary poles neutralize each other at the square marked by the arrow.

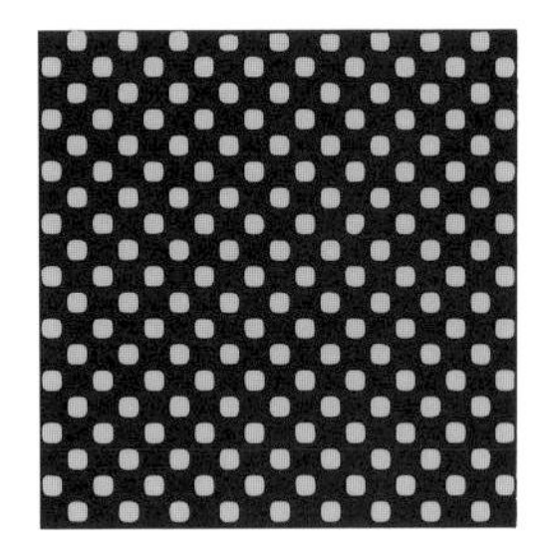

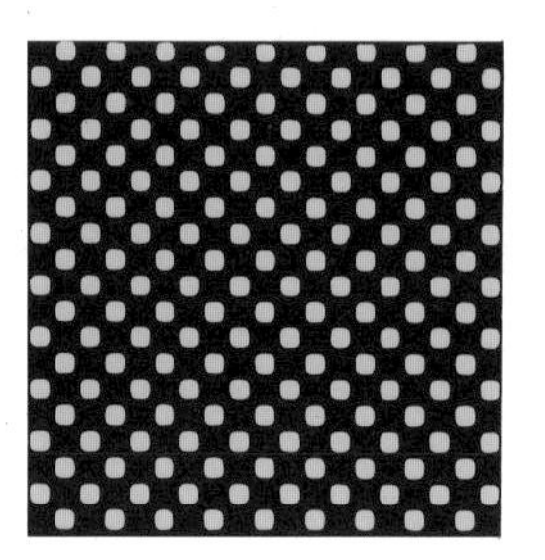

Simultaneous Effects:

Blue-Violet and Yellow-Green

A blue-violet and a yellow-green both appear gray.

From far enough away that the quantity and dimension of the dots can no longer be distinguished, the two squares above resemble the two squares surrounded respectively by yellow-green and blue-violet on the opposite page.

Maximum desaturation by a saturated color of the same hue.

The simultaneous complementary of blue-violet (citron yellow) mixes optically with desaturated blue-violet, and the simultaneous violet of yellow-green with desaturated yellow-green.

In consequence, the two small squares in the center appear gray.

The simultaneous complementary of blue-violet.

The simultaneous complementary of blue-violet is yellower than the optical complementary — that of yellow-green on the other hand is redder. (See the two semicircles on the opposite page.)

Successive image.

After staring at the black point for a sufficient length of time, you will perceive on a white sheet of paper four semicircles in which the lower half changes position with the upper half and vice versa.

Blue-violet + yellow-green = optical gray

A juxtaposition of 70% blue-violet dots with 30% yellow-green dots creates through optical combination a gray corresponding to the 70% black background (opposite page).

Maximum reinforcement of saturation (= purity and luminosity) by the complementary.

Here the simultaneous effects on page 105 are reversed. The simultaneous violet of the yellow-green background increases the brilliance of the desaturated blue-violet in the center. Next to that, the blue-violet background emphasizes the saturation of the central yellow-green.

Cleavage of gray into its two constituents, blue-violet and yellow-green.

The blue-violet background sets off the yellow-green elements. The yellow-green background activates a blue-violet tendency in the central square. The coefficient of simultaneous influence of 40% corresponds to the difference between the two colors on the opposite page. (Compare with the gradations of the scale on pages 102 and 103.)

Opposite page:

80 parts blue violet + 20 parts yellow-green = the desaturated blue-violet at left.

40 parts blue-violet + 60 parts yellow-green = the desaturated yellow-green at right.

The quantitative difference between these two colors is of 40 parts for the blue-violet and the yellow-green.

A

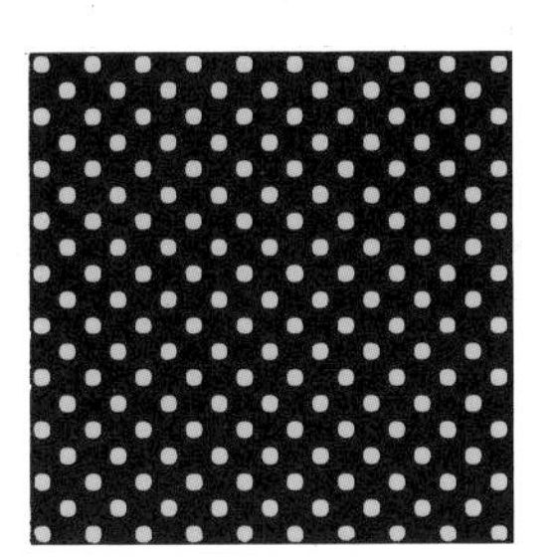

B

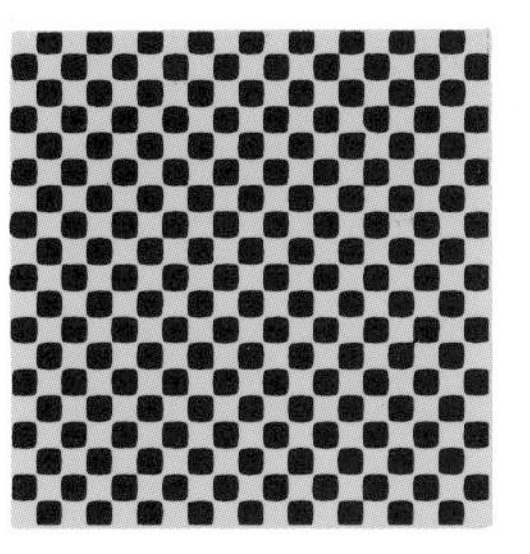

Psychological scale is different from physical scale.

A rotating disk of which one half is white and the other black will display a bright gray which is closer to white than to black.

A medium gray which to the observer appears to be halfway between white and black is created with only 20% to 25% white.

1. 0-13-26-38-49-59-68- [76] -83-89-94-100
2. 0- 5-10-15-20-30-40- [50] -60-70-80-100

1. The parts of black on a rotating disc correspond to

2. A sequence of numerical values in arithmetic progression, designating the percentages used in printing. In the boxed example, 76% black on a rotating disk (disk b) corresponds to 50% printed black (disc c).

This fact proves that psychological scale is not to be confused with physical scale. Various numerical formulas have been suggested to express the relation between the two scales: a logarithmic law (Delboeuf, after Fechner, 1872) and a parabolic formula (Plateau, 1873).[10] The Weber-Fechner law (in simplified and schematic form) states that the visual perception of an arithmetic progression (1-2-3-4, etc.) can be realized through a geometric physical progression (1-2-4-8, etc.). In *Interaction of Colors,*[11] Josef Albers deplores the fact that so few creators in the area of color are aware of this perceptual phenomenon. Scale 1 above shows a more progressive decrease in the parts of black from 0% to 100%.

Optical average: quantitative analogy between blue-yellow and white-black.

24 to 26 parts yellow, the brightest saturated color, when optically combined with 76 to 74 parts blue (100% cyan + 65% magenta) neutralize into the same medium gray (50% black — disc c) as 24 parts white combined with 76 parts black.

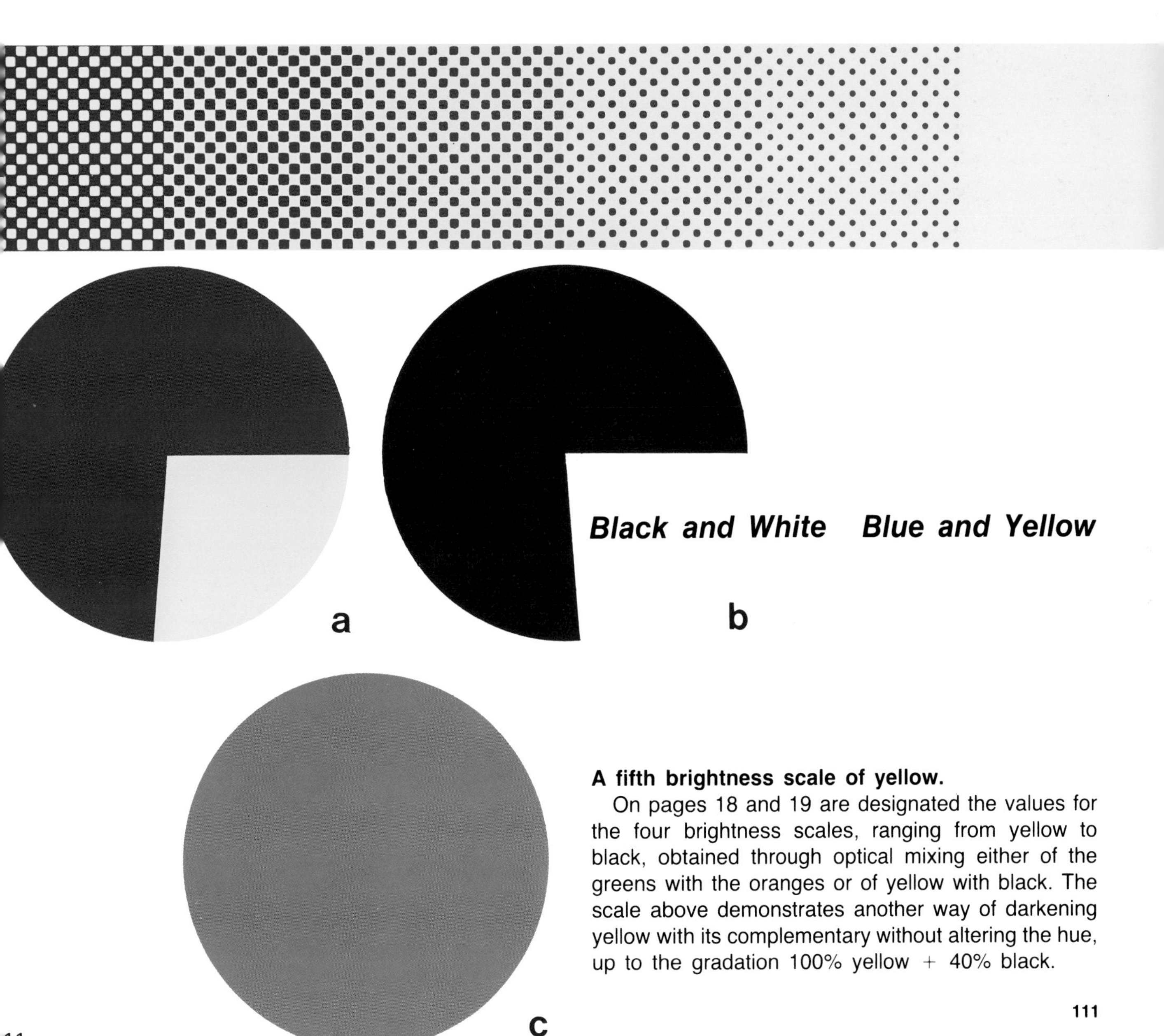

Black and White   Blue and Yellow

A fifth brightness scale of yellow.

On pages 18 and 19 are designated the values for the four brightness scales, ranging from yellow to black, obtained through optical mixing either of the greens with the oranges or of yellow with black. The scale above demonstrates another way of darkening yellow with its complementary without altering the hue, up to the gradation 100% yellow + 40% black.

Simultaneous Effect: Blue and Yellow

A blue and a yellow both appear gray.

From far enough away the two squares above resemble the two squares on the opposite page surrounded respectively by yellow and blue.

A desaturated yellow evokes a gray when associated with the simultaneous complementary of the saturated yellow background. In the same way blue-gray appears even grayer against a background of the same saturated blue.

Optical and simultaneous complementary of yellow.

The optical complementary of blue is yellow.

The simultaneous complementary, by contrast, is an oranged yellow.

The simultaneous complementary of the yellow is closer in appearance to the blue of the lower semicircle at right, a blue which more closely resembles blue-violet than the optical complementary. (See experiments on page 137.)

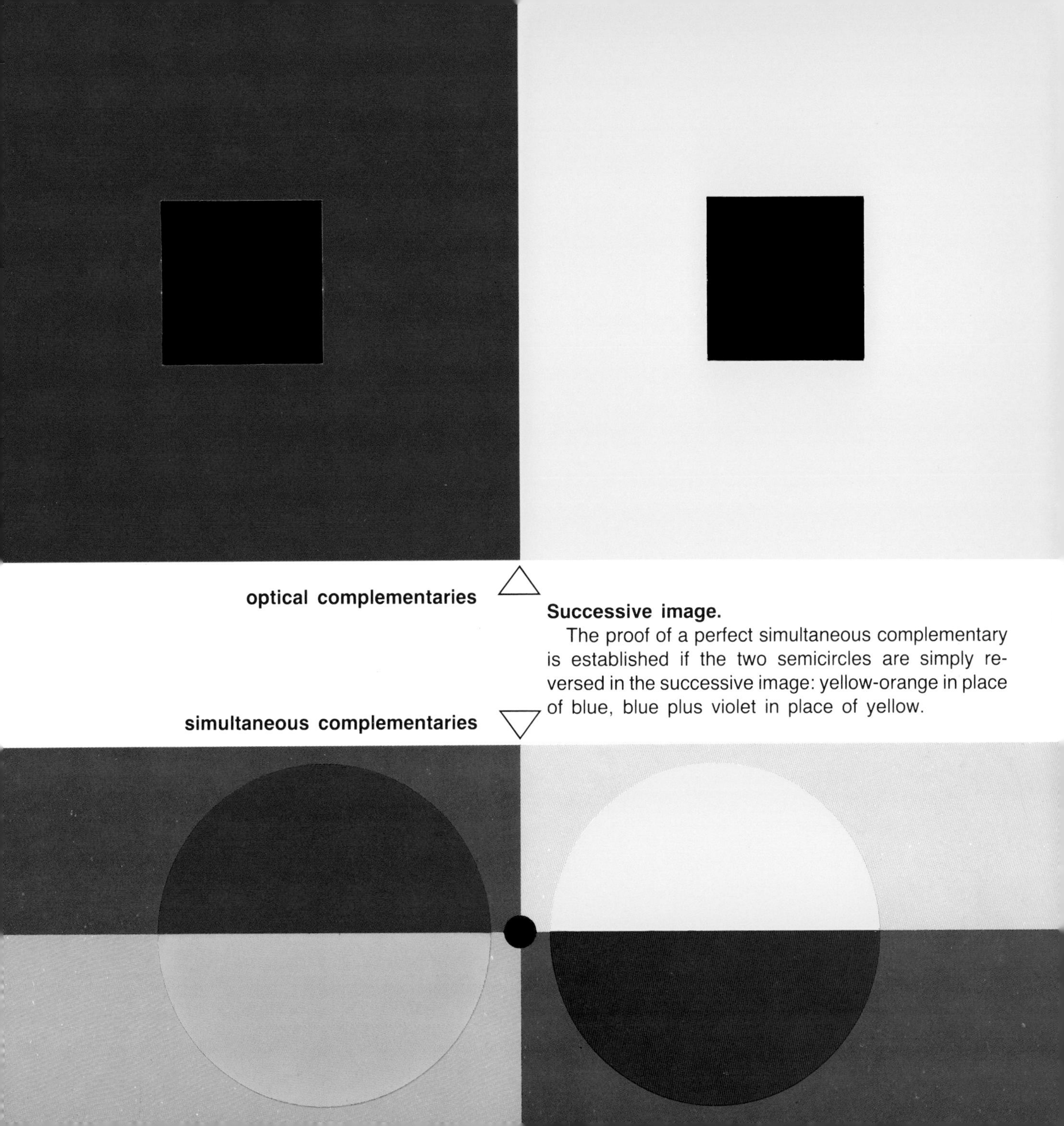

Successive image.

The proof of a perfect simultaneous complementary is established if the two semicircles are simply reversed in the successive image: yellow-orange in place of blue, blue plus violet in place of yellow.

Blue + yellow = optical gray.

The yellow dots when juxtaposed with the blue dots become indistinguishable against a medium gray background corresponding to disk c (page 111).

The dimension of the white dots of the background is identical to that of the yellow dots in the central squares. The black dots correspond exactly in size to the blue dots.

Verifying a perfect optical complementarity:

1. If the two squares above show a green tendency, only 5% of magenta is missing from the blue.

2. If there is an orange tendency, on the other hand, there is a slight surplus of magenta.

3. Finally, if a yellow tendency is observed, there is definite complementarity of hue, but it would be necessary to reduce the yellow dots slightly in order to create a neutral gray indistinguishable from the background.

Reinforcement of hue, saturation and brightness: The simultaneous effects on page 113 are reversed. The blue background, the optical complementary, accentuates the saturation and brightness of the yellow. The optically complementary yellow background, on the other hand, reinforces the brilliance of the blue and deepens its character.

A gray appears blue or yellow.
Gray takes on color through simultaneity:

The yellow background gives the gray a deeper appearance, with a blueish tendency, while the surrounding blue makes the gray brighter, and with a yellowish tendency. The coefficient of 50% indicates the impact of the simultaneous effects. It corresponds to the difference between the two colors on the opposite page:

90% blue + 10% yellow = the blue square (A)
40% blue + 60% yellow = the yellow square (B)

A

B

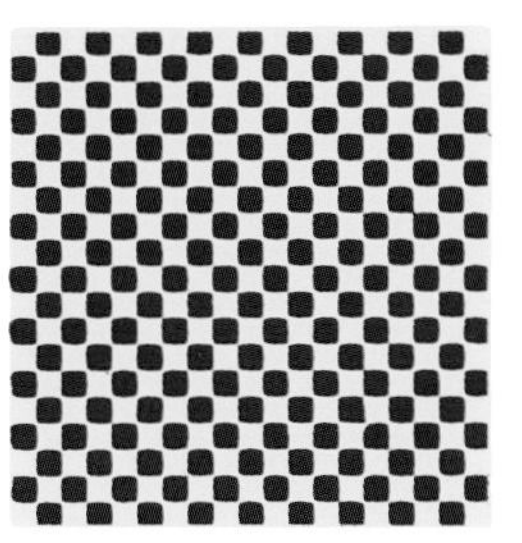

Successive hue and brightness of background.

The identical blue, seen against a white or a black background, differs not only in brightness but also in the hue of the successive image.

The successive contrast to blue against a white background reveals an orange desaturated by white which is slightly more yellow against a dark background; the same blue against a black background leads to the successive appearance of a more intense orange which is slightly more red against a bright background. Other examples are given on page 137.

The orange hue successive to blue against a gray background of the same brightness, intermediary between the two successive images above, corresponds to the simultaneous complementary.

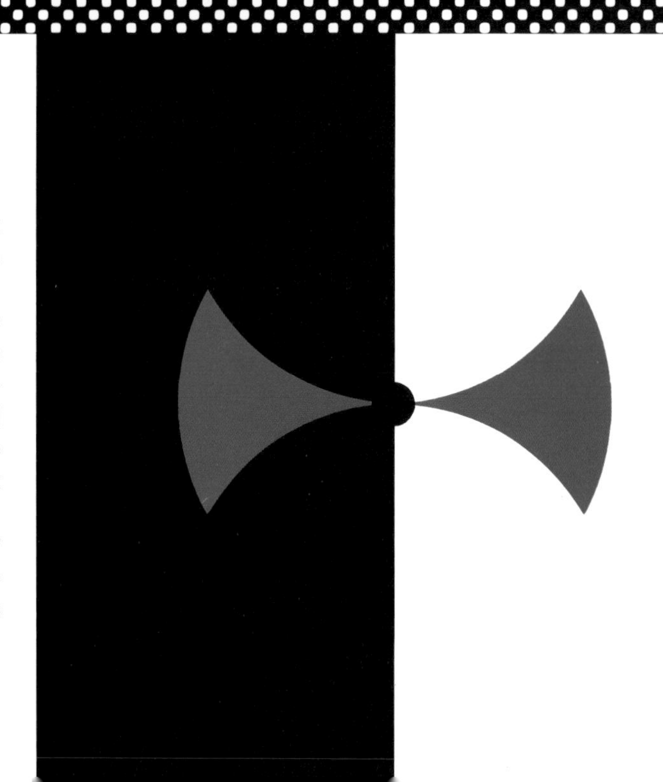

A bright gray and a dark gray appear identical.

If you stand far enough away so as to have no way to judge the size of the dots, perception must give way to abstract reasoning. The brightness of the medium gray on the half-page opposite now resembles the two squares in the center of the respective white and black backgrounds.

The surrounding black increases the luminosity of the dark gray area in the middle, while the surrounding white decreases the brightness of the bright gray central square.

Black + white = optical gray.
The white and black dots become detached from background above (60% black), which is too dark. The perceived gray resulting from their optical mixing blends more with the brightness of disc c on page 111 (50% black).

Maximum reinforcement of contrast of brightness.
The difference in brightness of the two central grays opposite is underscored by the contrast of the two backgrounds. The reason is that on the white background the deep gray darkens, while the surrounding black adds to the luminosity of the bright gray.

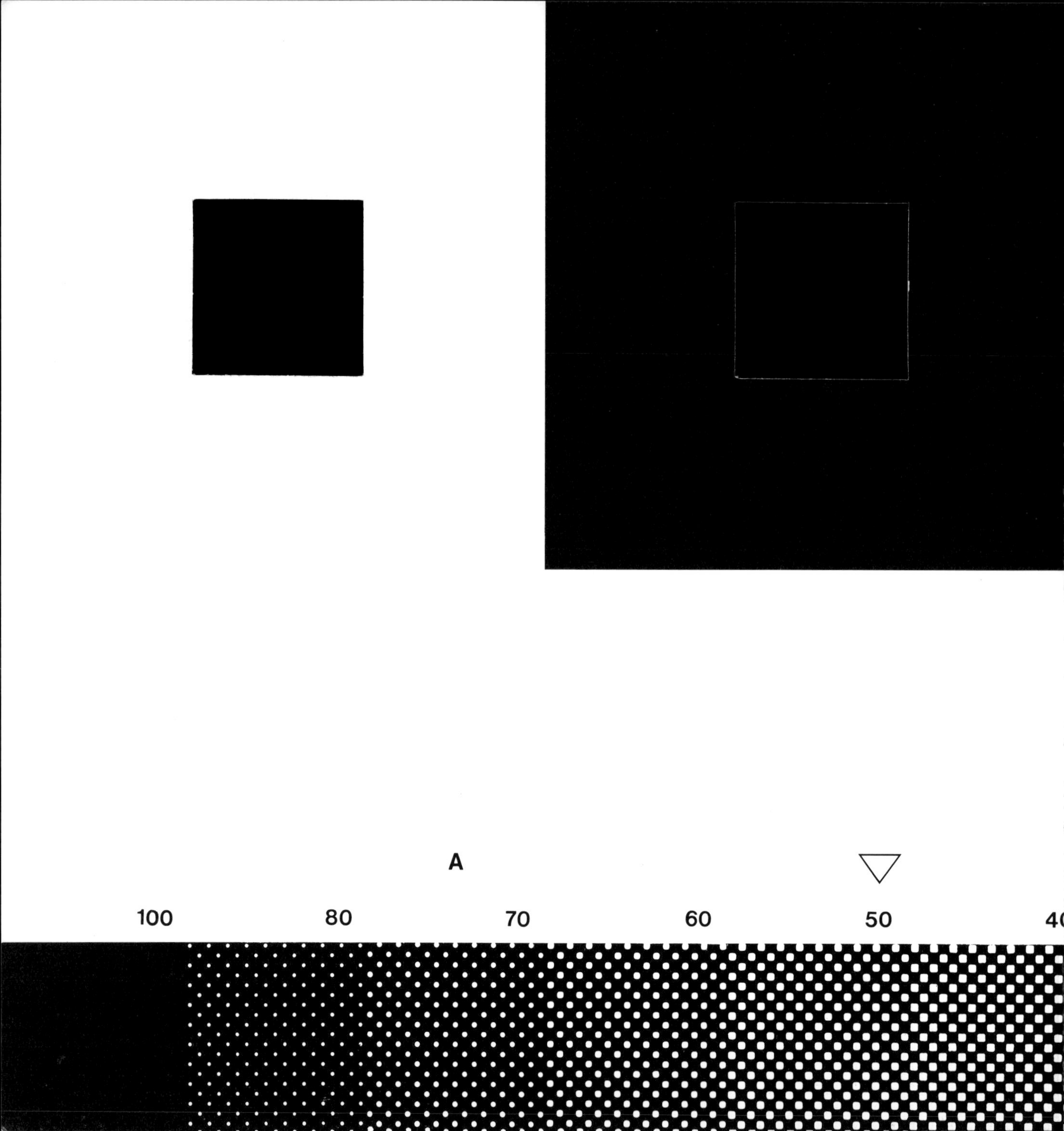
A
100
80
70
60
50
40

A

B

A medium gray appears bright or dark.

A medium gray (see the 50% gray on the scale) seems bright when surrounded by black; the same medium gray seems on the contrary a deep gray when surrounded by white. The coefficient of simultaneous influence on brightness is of 50%, corresponding to the difference between the two grays, A and B.

Under the same simultaneous conditions against a white background, the two grays above are perceived in their actual brightness.

Gray A consists of 70 to 80% black;

Gray B consists of: 20% to 30% black.

B

30 20 15 10 5 0

The Successive Image

Successive color is a phantom color, with no physical reality outside of an automatic psycho-physiological reaction produced by any visual stimulus stemming from the external world of luminous or material sources. In the simultaneous effect, the complementary coexists on the border of a colored surface; successive contrast is on the contrary a complementary projection extended in space/time and matches exactly the form and contours of the original chromatic stimulus. The successive complementary of simultaneous color is as strong in saturation and hue as the successive complementary of an objectively perceived physical color. In its pure state, in the double successive contrast, is is the complementary of the complementary of a colored background surrounding a neutral figure (preferably a gray of the same brightness as the background). For example, compare the successive images of disk c and d on pages 16 and 17, 30 and 31, and 44 and 45.

Every color we see can be considered as a positive imprint which possesses simultaneously a negative imprint. Stimulation by certain wavelengths and their complementary echo are two sides of the global visual potential which is always present even if most of the time we are completely unconscious of the phenomenon.

The reversed remnant interpreted by some authors simply as fatigue (a pause for the regeneration of the biochemical pigments of the cones and rods) seems to me rather to be part of an astute mnemonic process which makes it possible to relate a new situation to an old one.

The simultaneous or successive complementary reveals the functioning of the chromatic thresholds on the physiological level. Unfortunately it is a purely subjective experience which has to be translated into one of the synthetic systems referred to previously, in order to become visible to someone else and thus comparable and measurable.

Here let us recall a writer and poet profoundly interested in this visual phenomenon: Goethe,[12] who as early as 1810, in his treatise on color, remarked on the usefulness of the successive contrast in enabling us to precisely identify the physiological complementary.

How can the successive image be successfully evoked? There are some precautions which should be observed before beginning the experiment, in order to direct our attention solely to the space and the phenomenon in which we are interested. The book should preferably be placed against a neutral background. All the colored surfaces of the book should be covered by sheets of white paper, with the exception of the tints whose successive image we wish to see. Daylight without direct exposure to the sun is preferable to any artificial lighting. Now our glance should remain motionless for about 10 to 30 seconds, resting exactly on the black point. For any other color ensemble as well, it is essential to concentrate always and perseveringly on a central point. Next, the eyes should be raised abruptly toward a white surface fixed overhead; a single spot on this surface should be stared at until the successive image has completely formed. You must wait for the image to fade entirely before beginning another attempt.

The length of time the remnant persists may vary between a fraction of a second and several seconds. Do not lose patience if the successive colored apparitions are at first nonexistent, weak, or incomplete; as with anything, your performance will improve with practice.

In general, the same amount of time must be allowed for the successive image as was required for the initial stimulation. Nervousness, fatigue, or preoccupation may prevent you from seeing any trace of the successive phenomenon. Alertness, steady concentration, and mental relaxation, on the other hand, predispose in its favor.

The successive complementary is not immutable. It changes its character within very precise limits, relative to the spots of color making up the surroundings which enter into the visual field around the color we are fixing our gaze on. In other words, the successive image renders the complementary of an optical synthesis between an objective color (a luminous or pigmented source existing independently of the observer's eye) and the simultaneous complementary of its background.

There are four types. The adjacent colors will have an impact which can be:

1. Of opposition: two identical colors have a maximum successive dissimilarity when seen against two backgrounds of opposite successive effects.

2. Of reinforcement: when the background is composed of two colors whose simultaneous action leads in the same direction.

3. Neutral:

a. A gray background of the same brightness is neutral.

b. A background composed of the simultaneous complementary is neutral.

c. Also neutral, but very complex in terms of evaluating quantities, is the presence of two or more colors with an opposite simultaneous impact.

d. A gray surrounded by two complementaries in the proportions of optical mixing does not undergo any change.

4. Desaturating: a background whose successive complementary is close to the simultaneous complementary of the color gazed at has a desaturating effect.

Concrete demonstrations will be found on pages 128 and 129, using the examples of the transformation of the central yellow-green.

Colors standing out against a white background will not only appear brighter in the successive image, but also undergo a change of hue. They resemble the optical complementaries. Colors gazed at against a black background, especially under an opposite simultaneous influence, appear more intense or darker, and in terms of hue recall the subtractive complementaries. (For further details, see pages 136 and 137.)

As for the differences in size of the successive impressions in relation to their distance[13] the following observations can be made: if your successive projection falls on a surface farther from you than the printed area, it will appear larger than the printed area. If conversely it falls on a surface nearer than the source-image, the successive response will be smaller. This is known as the Emmert law. It specifies that the successive image approximately doubles in size if the surface on which it appears is twice as far away.

The constancy principle proves that it is not the retinal image alone which determines the size, but that the brain makes adjustments in order to facilitate the interpretation of the change.

It is virtually impossible to render the successive image's intense diaphanous luminosity with material colors such as printing inks. The comparative examples shown here can claim only to furnish the tendency of the hue. Furthermore, from one day to another fluctuations around an average are often observed, depending on changes in lighting, fatigue, availability, etc.

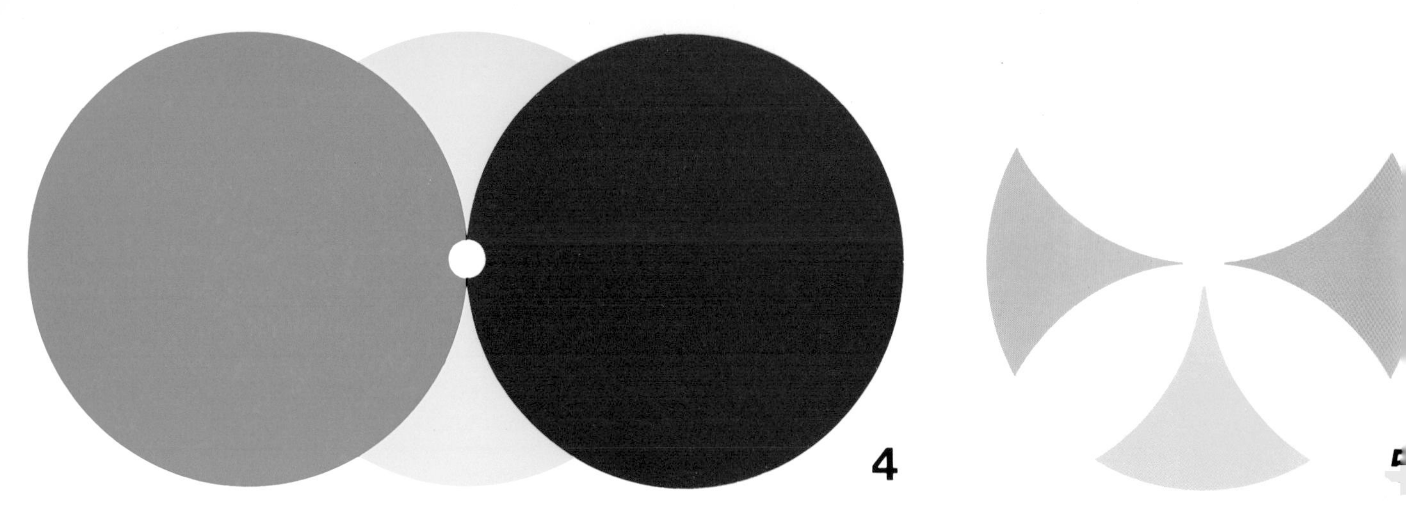

The successive poles of blue-violet.
Having first covered all the colors except for group 1, fix your attention on the black point in the way previously described. By abruptly shifting your glance upward onto the white screen, the successive tendencies of the blue-violet (the two horizontal curvilinear triangles of Figure 5) will be perceived — a green at left, an orange at right.

Two neutral backgrounds.
You can proceed in the same fashion with the gray semicircle, juxtaposed with the semicircle of the simultaneous complementary (2). The successive image of the blue-violet yields this time a slightly greenish yellow (see the vertical element in Figure 5).

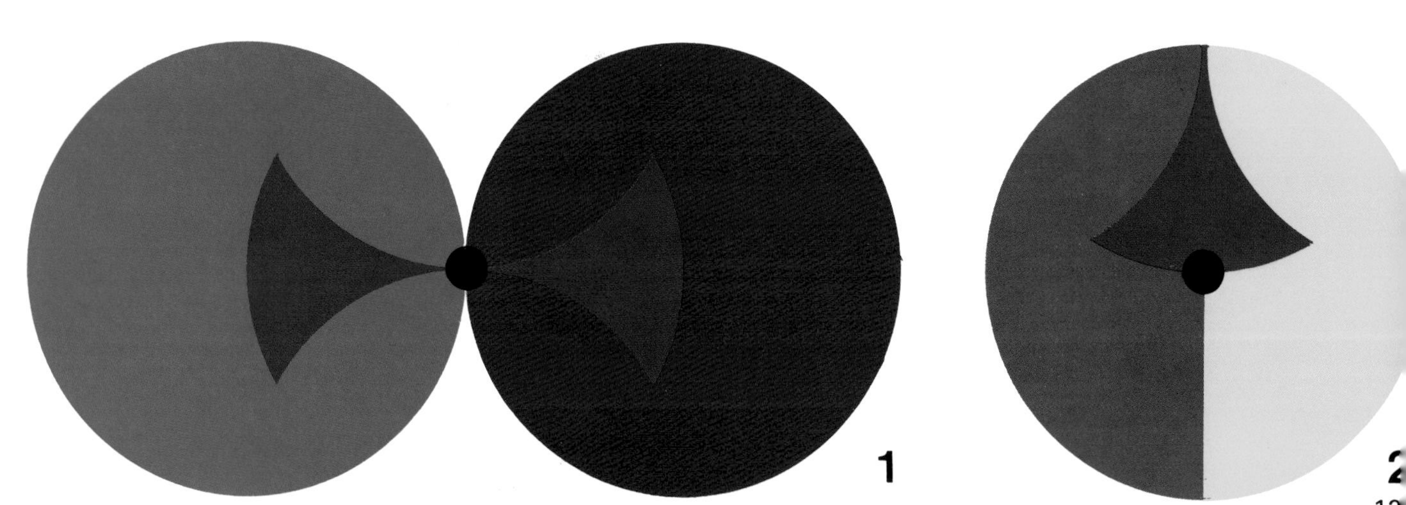

Successive image of the six basic colors.

What will be the appearance of the successive image of the three additive primaries, red-orange, green, and blue-violet, in a unitary visual field? We see the successive contrasts, corresponding approximately to Figure 4 on the opposite page, standing out clearly against the white background. The cyan is slightly greener and the magenta a bit more violet. (If adapted to a white background, there is a reduction in the optical direction.) The color in the middle is unequivocally a yellow. In this situation of opposition, the successive complementary takes on the aspect of the subtractive complementary. Red-orange and green, both optically and through addition, create yellow. The successive complementary of blue-violet probably bears the imprint of this visual stimulation of the primaries.

Reversing the experiment, the successive impression of the three subtractive primaries of Figure 4 reveals hues which are notably close to Figure 3, adapted to a black background.

Mixed successive contrast.

Mixed contrast makes its appearance when the successive image falls on a colored background with which it interacts to produce a new optically combined color which is the synthesis of the successive image and the color of the background.

Examples:

When the successive image of Figure 3 below falls on a colored background, the synthesis of this projection and the background reveals the following colors from left to right:

In combination with a yellow background, green, yellow, red-orange.

In combination with a magenta background: violet, red-orange, magenta.

Successive projection in combination with a cyan background: cyan, green, violet.

Curiously, it is possible sometimes to reconstitute the three primaries (cyan, yellow, magenta) which seem to float in front of the cyan background with which they do not combine.

The yellow successive complementaries against a yellow background, magenta against magenta, cyan against cyan stand out by virtue of a luminosity greater than that of the background.

3

Successive Images of Yellow-Green

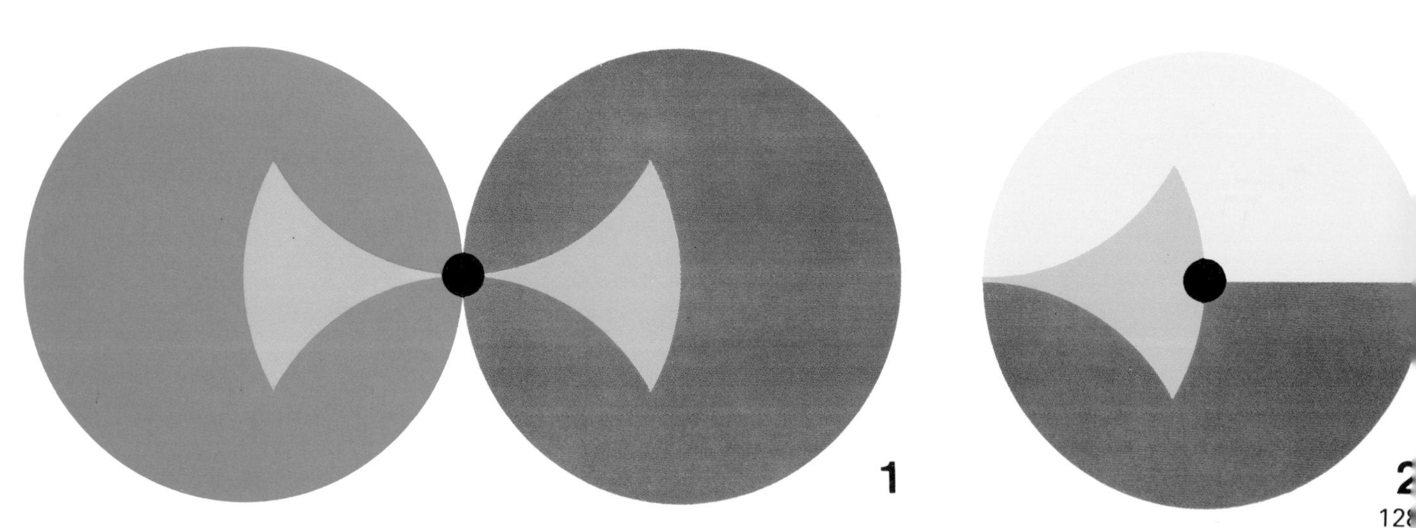

1. *Opposition:* the central yellow-green changes into successive blue when seen against a blue-green background, and into successive magenta when seen against an orange background. For yellow-green, all green, blue-green, and blue backgrounds are in opposition to yellow, orange, red, and magenta backgrounds.

2. *Reinforcement:* a background consisting of two colors (yellow and orange) whose simultaneous activity tends predominantly toward magenta.

3a and b. *Neutrality:* the influence of a background consisting half of gray and half of the simultaneous complementary of yellow-green is neutral (3a). The opposite situation, in which there are equal parts of orange and blue-green, has similarly neutral effects (3b).

4. *Desaturation:* the successive image of yellow-green when we gaze at it against a yellow background is an extremely desaturated red, since blue-violet is the optical complementary of yellow-green (blue-violet + yellow-green = gray; see page 103). The simultaneous complementary of yellow is very close to blue-violet.

5. Hues which approximately reproduce the successive images of yellow-green: horizontal, Figure 1; vertical, Figures 3a and 3b.

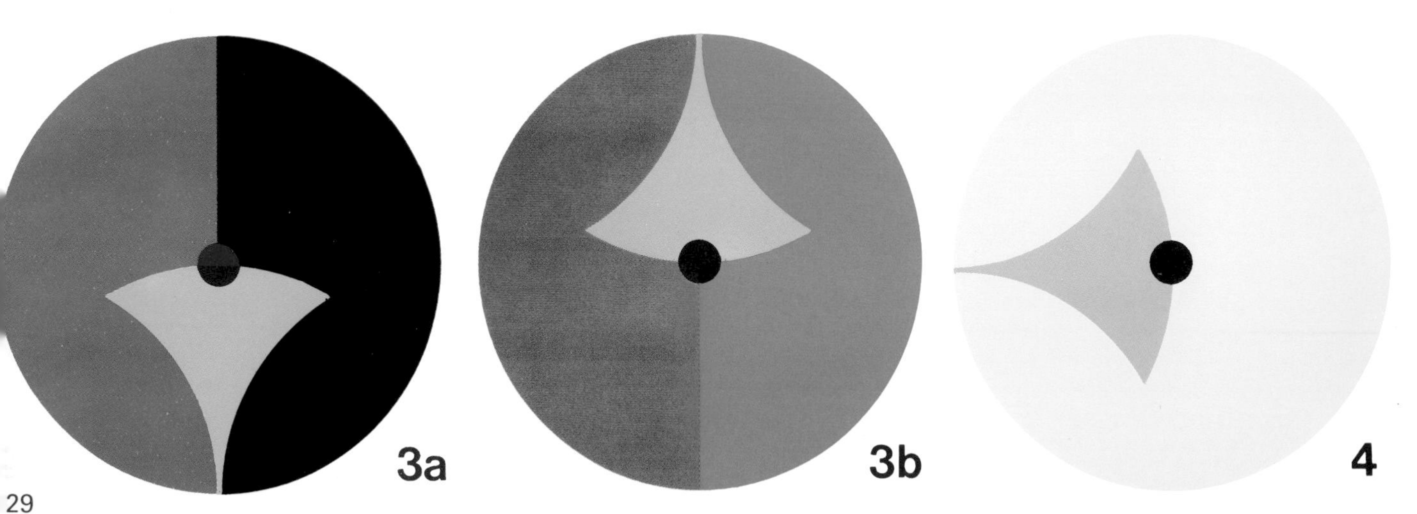

Successive Images of Blue

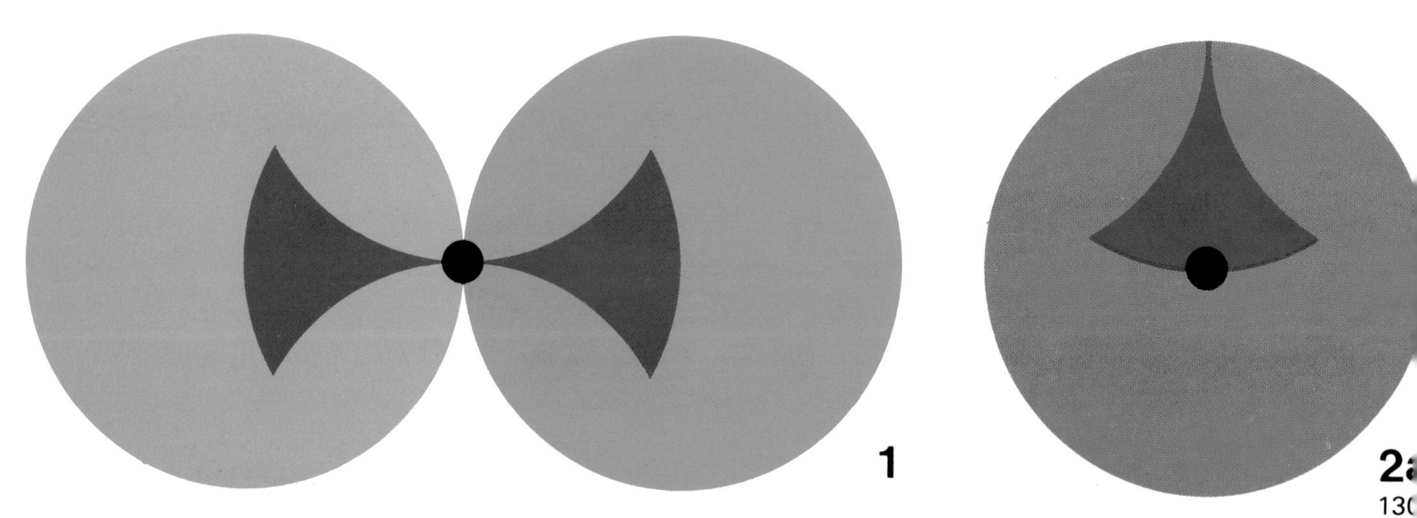

Successive poles of blue.

The color gazed at, identical for 1, 2a, 2b, and 3 is a blue (100% cyan + 65% magenta) midway between cyan and blue-violet.

1. Of the two backgrounds at left, both of which emphasize the yellow tendency successive to blue, it is the cyan background which brings out a more desaturated yellow, with a suggestion of green.

2. Two neutral situations:

a. With a background of the simultaneous complementary of blue;

b. With an optical mix of two backgrounds with simultaneous tendencies antagonist to blue.

3. In opposition to cyan, blue-green, and green backgrounds with the yellows and greenish yellows successive to blue (Figure 1), the red, magenta, and red-violet backgrounds evoke successive oranges, as Figure 3 demonstrates.

4. Successive images:

Horizontal triangle at left for Figure 1.
Horizontal triangle at right for Figure 3.
Vertical triangle for Figures 2a and 2b.

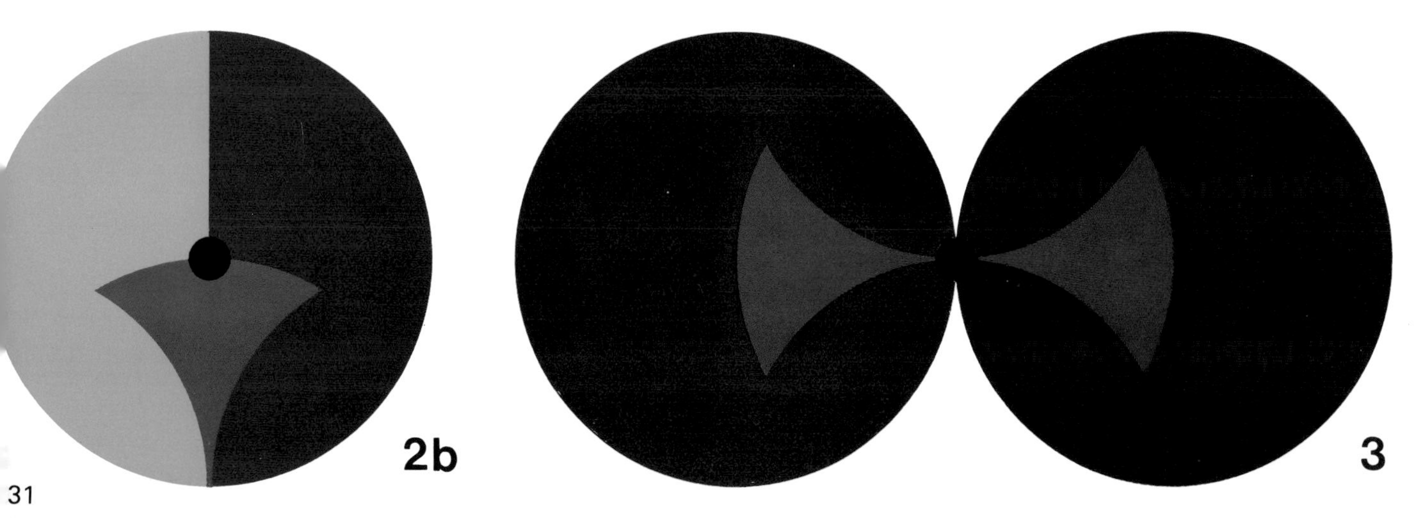

Successive Images

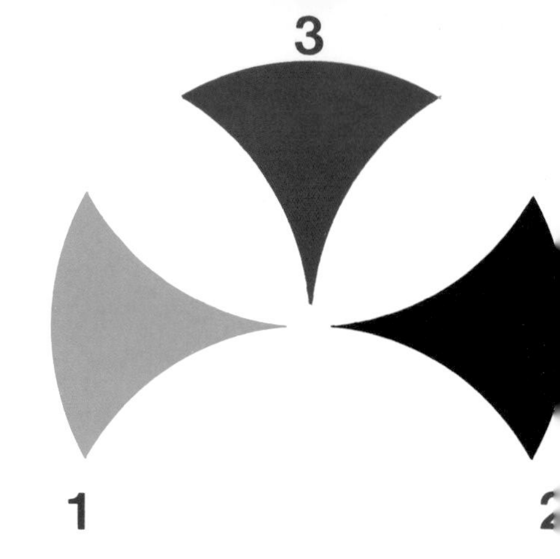

Successive poles of orange.

The same orange is surrounded by two colors of the same vicinity but antagonistic in terms of their simultaneous effects:

1. the yellow background at left evokes a successive cyan;

2. the orange background at right evokes a successive red-violet.

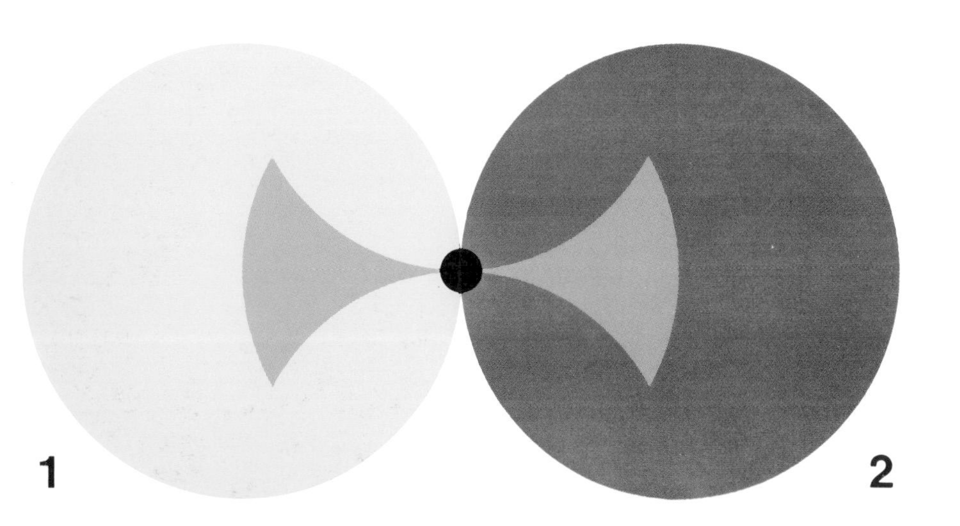

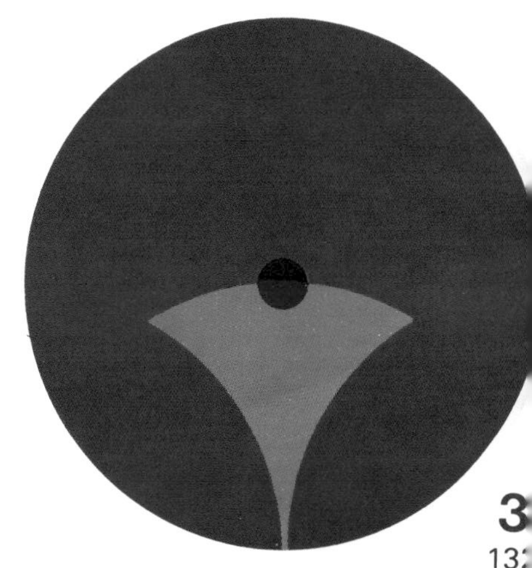

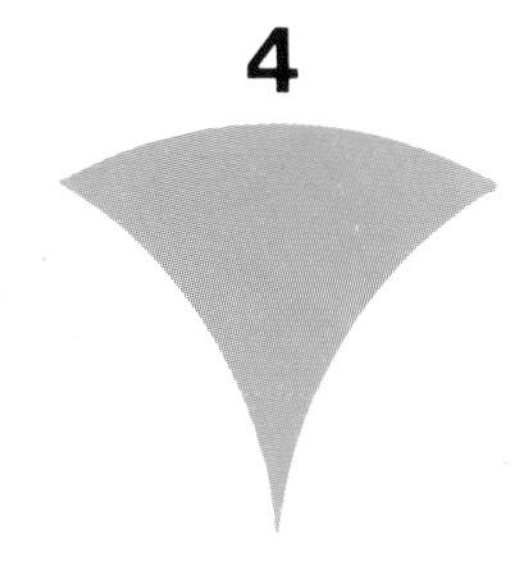

Double successive contrast:

3. Between these two poles is situated the simultaneous complementary, which is here, in the successive image, the complementary of the complementary of the blue circle at bottom. The neutral gray of the same brightness as the blue supports the transmutation of a simultaneous impression into a successive impression.

Two complementaries are neutralized in the successive image:

4. The simultaneous effects of about 30 parts of orange with 70 parts of blue (a pair close to the subtractive complementaries) mix optically in such a way that the medium gray triangle remains neutral in the successive image. Despite a fluctuation toward orange along the border with blue and toward blue along the gray/orange border, the successive image of the gray stabilizes around a medium gray. This neutrality results with most complementaries of blue, from the subtractive complementary orange to the optical complementary yellow by way of the simultaneous complementary orange-yellow.

Surrounded by a large white background, the blue/yellow pair of Figure B, page 16, displays the expression of optical complementarity in the successive image. Surrounded by a large black background, the blue/orange pair of circle 4 opposite displays the expression of subtractive complementarity in the successive image.

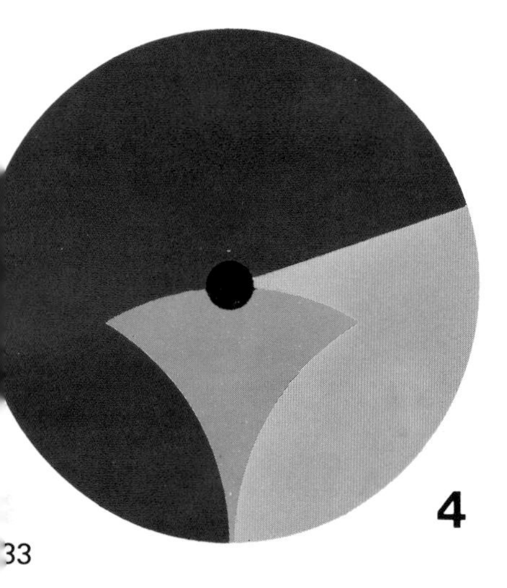

Successive Images of Nonsaturated Colors

Below we see three colors of the same hue against a white neutral background, of which only the one in the middle is 100% saturated. At left we have magenta + black, at right magenta + white.

The successive effects reveal a whitish desaturated green for the dark magenta and a dark green for the bright magenta. A pure and luminous green emerges between them, reproducing the saturation of the magenta in the center.

The successive image of a color desaturated by either white, black, gray, or the complementary, and gazed at against backgrounds of different colors, not only changes in light intensity as compared with the saturated color; an enlargement of the differences in hue, in comparison with the summary on pages 138 and 139, can also be observed. The complementary of the background compensates for the desaturation by yielding a successive image of the purest and most luminous desaturated color. Three examples follow:

On page 28, apply the transparent yellow sheet. Focus your attention on the black dot. The successive image of the desaturated yellow in the center of the green is a blue close to cyan. The same desaturated yellow in the center of the red-orange changes into a successive hue close to magenta. The green and red-orange backgrounds, in a simultaneous action, decompose the complementary of the desaturated yellow into, approximately, a successive cyan and magenta.

Opening to page 42, apply the magenta sheet. Gaze at the group in the middle: the red-orange background juxtaposed with the blue-violet background. The simultaneous impact of this pair on the central desaturated magenta will be such that in the successive image it appears almost yellow in the center of the cyan and cyan in the center of the yellow (a decomposition of its complementary, green).

On page 56, apply the transparent cyan sheet. An identical desaturated cyan is in the center of the green and the blue-violet backgrounds. After fixing your eyes on the border of the two backgrounds, and then projecting onto a white screen, the successive result is as follows: a breaking down of the complementary of the desaturated cyan into its two components, close to magenta at right and close to yellow at left.

Experimenting with the Successive Image

We have seen in the last chapter that the successive image depends on the hue and brightness immediately surrounding the color gazed at. You can compare your own color impressions with the summary on pages 138 and 139 by applying the color disks to surfaces of different hues, or by cutting out the curvilinear triangles on page 147 and following the instructions.

After cutting out the red-orange, green, blue-violet, and white curvilinear triangles, here is the proper method for pages 28, 56, and 43:

1. Place the curvilinear triangles along the common border of the two backgrounds, so that they touch vertically at the center of the black dot. You can now hide the areas containing the small squares in the center of the background with your hands or a sheet of paper.

2. After clearly determining the successive image, turn the cut-out shape so that it forms a diagonal with the border. (The small squares remain hidden.)

3. Stepping further back, your glance, while remaining fixed on the black dot, can now embrace the whole of both backgrounds. A comparison between the small central squares and the curvilinear triangles becomes possible.

You can begin by using the colored shapes, and then the white shape, covering it with a transparent sheet of the corresponding color.

Page 28: blue-violet and yellow

1. Cyan and magenta act together on the vertical blue-violet and successively create the third successive primary, yellow. White + the transparent yellow: the situation is reversed. (See also page 22.)

2. Diagonal: successive images — orange, gazed at beforehand against a magenta background; yellow-green, gazed at against a cyan background. For yellow: blue, against a green background; red-violet, against a red-orange background.

3. Comparison of simultaneous effects on saturated blue-violet with desaturated blue-violet: desaturated blue-violet is closer to red-orange in the center of the successive green, and closer to green in the center of the successive red-orange.

For yellow, see page 134.

Page 42: green and magenta

1. Cyan and yellow are equally responsible for the magenta successive to the green positioned vertically along their common border. White + the transparent magenta: a successive green, subtractive complementary.

2. Diagonal: successive images — red gazed at beforehand against a yellow background; violet gazed at against a cyan background. For magenta: successively, a yellow-green surrounded by cyan; a blue-green surrounded by yellow.

3. Comparisons: the successive desaturated green becomes definitely orange in the center of a blue-violet, and blue-violet in the center of the orange. For magenta, see page 134.

Page 56: red-orange and cyan

1. The red-orange in the center undergoes, quantitatively and qualitatively, the same simultaneous effects as yellow and its antagonist hue, magenta. Successive appearance: cyan. White + the transparent cyan: the successive image reveals a red-orange in the center of the yellow and the magenta.

2. Diagonal: successive to the saturated red-orange gazed at against a magenta background, blue is perceived; for saturated cyan: red against a blue-violet background and orange against a green background.

3. Comparisons: the successive complementary of desaturated red-orange is divided between the two components of cyan, green in the center of the successive blue-violet, blue-violet in the center of the successive green. For cyan, see page 134.

Optical mixing of two consecutive images in a successive impression.

It is fascinating to observe, in the successive image, the neutralization of two complementaries occupying the same space but gazed at at different times.

Example:

Surround the disk on page 22 with a large black background. Gaze at the central dot for at least 10 seconds, without shifting your glance; then add the transparent yellow disk on top of it, and continue gazing for about 5 to 10 seconds.

The successive image will evoke a red semicircle (halfway between magenta and red-orange) alongside a blue-green semicircle (halfway between cyan and green) and a central gray figure, a neutralization of yellow + black with blue-violet + white.

The two consecutive images, gazed at in the same space and dividing between each other a gazing period not exceeding 20 to 30 seconds in all, have mixed optically in the eye, creating a successive, intermediary, unique impression.

Successive Complementary: Optical or Subtractive?

Can the successive complementary correspond to the optical or the subtractive complementary? Surprisingly, the hue approaches one or the other depending on the total disposition of energy surrounding the complementary pair which is gazed at.

If a white visual field reflects all the chromatic energies, it is the entire retina, in collaboration with the brain, which brings about a reduction of the thresholds in the direction of optical synthesis (reduction of magenta, augmentation of cyan, greater or lesser quantity of yellow).

If the visual field surrounds the optical complementary pair with black on all sides, there is an absorption of all the chromatic energies by the field, and the total visual potential begins to function (cyan, magenta, and yellow participate with the same intensity in the direction of the subtractive process).

Conditions for experimentation:

A large black (white) background, about 50 cm. by 100 cm. for example, should cover the entire visual field. It is important to position yourself above the example to be studied in such a way that even the peripheral regions of the eye are saturated with white or black. To be sure of a perfect optical complementarity, the pairs must be selected by verifying beforehand, with the aid of the rotating disks, the neutrality of the grays to be synthesized. Now, combine two complementary semi-circles on a single disk which is placed in the center of the white or black background. A white sheet is juxtaposed with the black background, creating a projection screen.

Among the examples cited in the summary table, here are the three most spectacular cases:

Yellow and blue.

The pair consisting of blue and its exact optical complementary yellow together make up a disk 6 cm. in diameter. Gazed at in the center of the white background, the successive image projected onto the nearby white yields an optically complementary luminous blue and a yellow brighter than the saturated yellow gazed at.

Gazed at in the center of the black background, the successive image shows a blue-violet, recalling the blue-violet which is subtractive complementary of the yellow, and an orange close to the subtractive complementary of the blue.

Cyan and orange.

When optical complementary pair cyan and orange are gazed at in the center of the white, there is a successive inversion of hue, but brighter and more luminous than the objective pair.

In the center of the black, the successive image of the cyan reminds us in its intensity of the subtractive complementary red-orange, while the orange projects a successive blue whose percentage of magenta corresponds closely to the percentage by which the orange falls short of 100% saturation.

Red and blue-green.

The optical complementary pair red and blue-green is exceptional. In this region, optical and subtractive complementaries coincide. Experimentation against neutral backgrounds confirms this fact: the successive hues undergo no transformation. The image of the pair is simply reversed, except that it is darker when gazed at against a black background and brighter against a white background.

The blue-violet, optical complementary of the yellow-green (30% cyan + 100% yellow), gazed at against a black background, yields a successive yellow; the blue-violet, optical complementary of the yellow-green (40% cyan + 100% yellow), on the other hand, yields a greenish yellow.

Summary Table of the Successive Images

The summary table on pages 138 and 139 represents the principal tendencies of transformation of the successive images (3) of a saturated color (1) which has undergone the simultaneous influences of different colored backgrounds (2). The successive image of a color (1) gazed at against a black background is darker and close to the subtractive complementary (4). It is brighter and close to the optical complementary when gazed at against a white background (5).

Backgrounds consisting of neighboring hues are responsible for the strongest alterations. Between these two extremes are the intermediary successive colors with, in the center, the background of the simultaneous complementary (an intermediary hue between the optical and the subtractive complementary) which alone can restore a successive image identical to the background.

The colors in parentheses refer to the fact that desaturated colors under a more powerful simultaneous influence go beyond the limitations of this summary's framework.

Color gazed at	Colored backgrounds	Successive Images	Black background Subtractive complementary (dark)	White background Optical complementary (white)
		(yellow)		
magenta	red	yellow-green		
(desaturated	orange			
magenta)	yellow	green		
	green		100% cyan + 100% yellow	100% cyan + 70% yellow
	cyan			
	blue	blue-green		
	violet	(cyan)		
red-violet	magenta	greenish yellow		
100% magenta +	orange			
30% cyan	yellow		100% yellow + 70% cyan	100% yellow + 100% cyan
	green	green		
	cyan			
	blue-violet	green		
blue-violet	magenta	orange		
100% cyan +	orange			
100% magenta	yellow		100% yellow	100% yellow + 40% cyan
	yellow-green	yellow-green		
	green			
	cyan	green		
blue	violet	red-orange		
100% cyan +	red			
65% magenta	orange		100% yellow + 35% magenta	100% yellow
	oranged yellow	oranged yellow		
	yellow			
	green			
	cyan	yellow-green		
		(magenta)		
cyan	blue	red		
(desaturated	violet			
cyan)	red		100% yellow + 100% magenta	100% yellow + 50% magenta
	orange	orange		
	yellow			
	green	oranged yellow		
		(yellow)		
blue-green	blue	magenta		
100% cyan +	violet		100% magenta + 50% yellow	100% magenta + 50% yellow
50% yellow	red	red		
	orange			
	yellow			
	green	orange		

Color gazed at	Colored backgrounds	Successive images	Black background Subtractive complementary (dark)	White background Optical complementary (bright)
green	yellow	orange		
100% cyan +	orange			
100% yellow	red	magenta	100% magenta	100% magenta + 30% cyan
	red-violet	red-violet		
	violet			
	blue			
	cyan	violet		
yellow-green	yellow	red		
100% yellow +	red			
40% cyan	violet	violet	100% magenta + 60% cyan	100% magenta + 100% cyan
	blue			
	green	blue		
		(magenta)		
yellow	orange	red-violet		
(desaturated	red			
yellow)	violet		100% magenta + 100% cyan	65% magenta + 100% cyan
	blue	blue		
	cyan			
	green			
	yellow-green	blue		
		(cyan)		
orange	red	violet		
100% yellow +	violet			
50% magenta	cyanated blue	cyanated blue	50% magenta + 100% cyan	100% cyan
	green			
	yellow	blue-green		
red-orange	magenta	blue		
100% magenta +	violet			
100% yellow	blue			
	cyanated green	cyanated green	100% cyan	100% cyan + 35% yellow
	green			
	yellow			
	orange	green		
red	red-violet	blue-cyan		
100% magenta +	blue-violet			
50% yellow	blue			
	blue-green	blue-green	100% cyan + 50% yellow	100% cyan + 50% yellow
	green			
	yellow			
	orange	green		

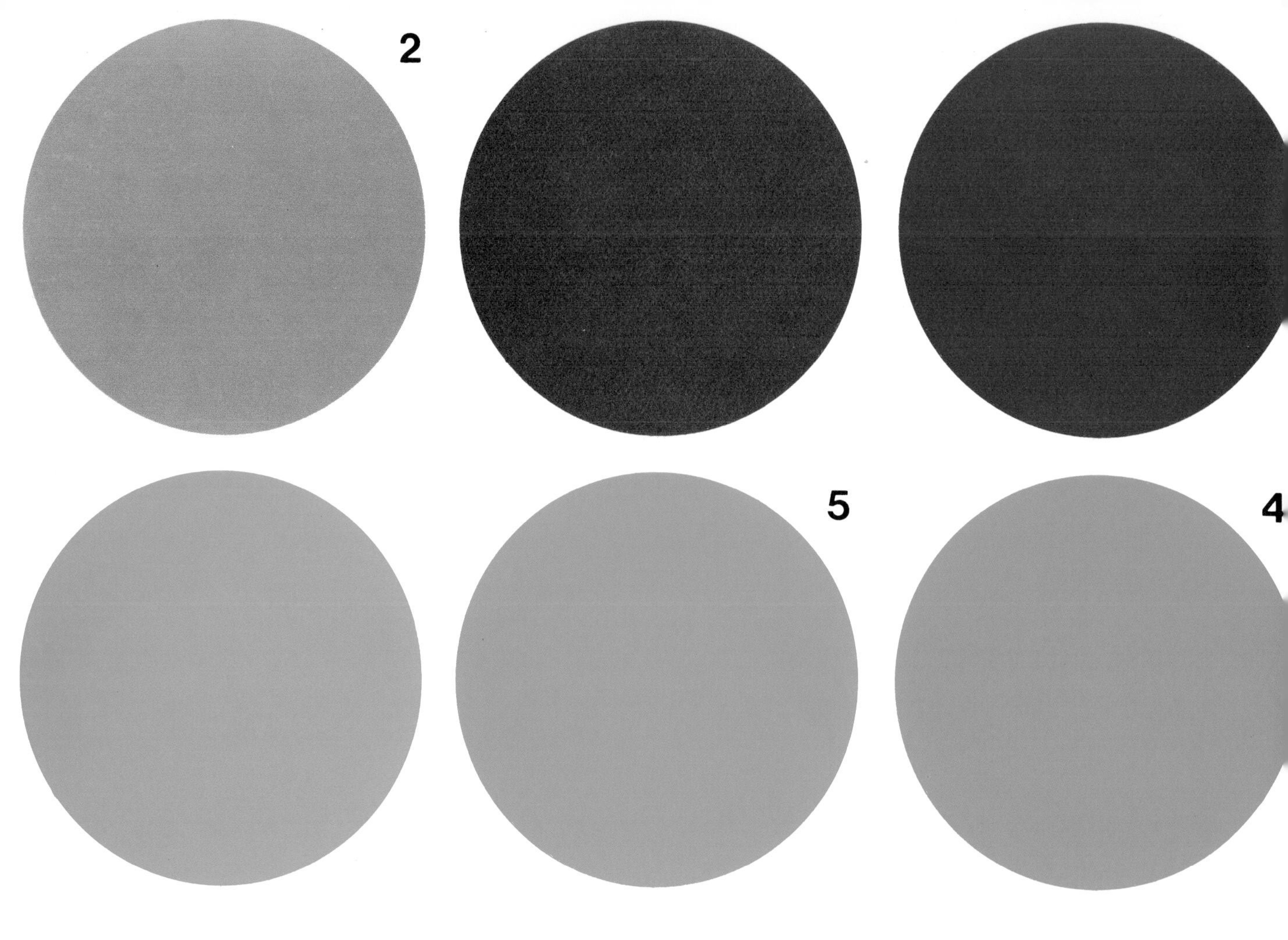

Experimenting with Rotating Disks

To experiment on your own, cut out the disks on pages 144, 145, 147, and 149, which represent the 12 principal colors + black and white which we have been studying throughout this book.

For optical combination, it is easy to adapt a motor, of the electric drill variety, to the requirements of rotating disks. All you have to do is mount, in place of the drill, an axis supporting a round plate with a minimum of 6 centimeters in diameter. For instance, paper disks taken from a sanding machine can be used for this purpose. This support-disk must be sufficiently stiff and perfectly centered. A speed of 3,000 rpm is generally suitable for the fusion of the colored sectors.

Having beforehand cut with a razor blade a slit of the width of the radius in the colored disks, it will then be possible to thread them through, in pairs or even with three colors, one on top of the other.

With the help of the graduated circle on page 142, you can measure the ratios of the colors in exact percentages.

To cut out: the primaries (page 147):

1. green (100% cyan + 100% yellow)
2. cyan
3. blue-violet (100% cyan + 100% magenta)

Primaries (page 149):

4. magenta
5. red-orange (100% magenta + 100% yellow)
6. yellow

with their optical complementaries (pages 144–45):

1. red-violet (100% magenta + 30% cyan)
2. orange (100% yellow + 50% magenta)
3. yellow-green (100% yellow + 40% cyan)
4. green (100% cyan + 70% yellow)
5. blue-green (100% cyan + 35% yellow)
6. blue (100% cyan + 65% magenta)

Even minimal differences in the working conditions of the reproduction techniques — to name only temperature and surrounding humidity, etc. — can cause variations of 5 to 10% in the desired norm. A difference of that magnitude is enough to falsify the ratios of complementarity. For this reason, each optical complementary is printed, on pages 144 and 145, in three versions with 5 to 10% less (top of page) and 5 to 10% more (bottom of page). In principle, the optimal values will be found in the middle of the page. In the hypothetical case of a lapse in the reproduction, there are three different options for selecting the true optical complementary.

The pair primary green/red-violet (1) does not correspond to the exceptional pair red-violet + green which neutralizes in equal ratio, as described on pages 94 through 101. This one is composed of green (100% yellow + 85% cyan) and red-violet (100% magenta + 40% cyan).

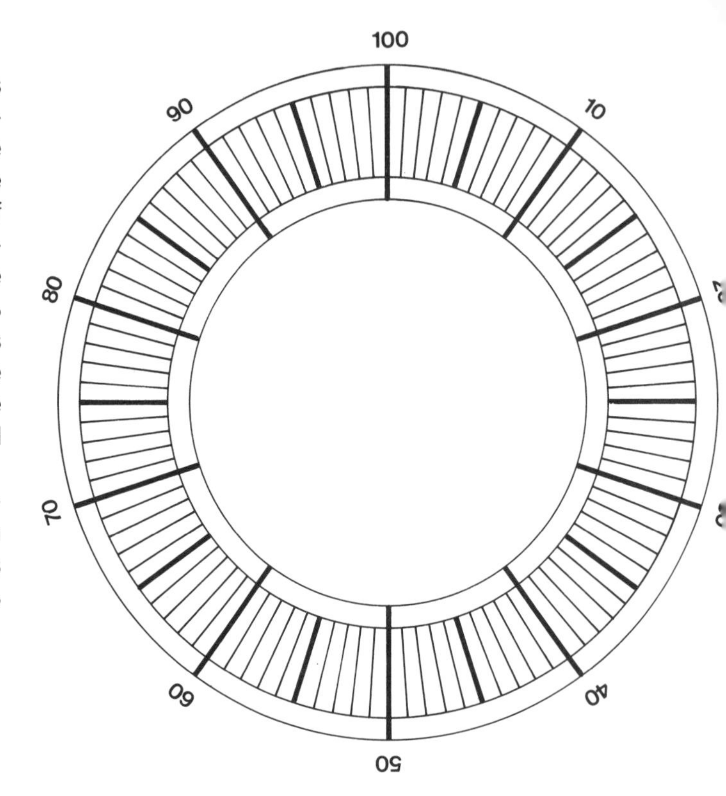

Measuring Percentages with a Graduated Circle

A circle graduated from 0% to 100% permits a mathematical evaluation of the ratios of the different colors in question. You can augment or diminish at will the section of one color in relation to another and in this way reconstitute the various stages of all possible scales. Thus, you can verify the premises of this book, by seeking out the optical average of the six fundamental scales:

1. Of a subtractive primary, by way of an average brightened by white, to another subtractive primary:

from magenta to cyan; from yellow to cyan; from yellow to magenta.

2. Of a secondary color, by way of an average darkened by black, to another secondary:

from red-orange to green; from red-orange to blue-violet; from green to blue-violet.

3. By creating scales of brightness:

from white to black; from white to a saturated color; from a saturated color to black.

4. By selecting the optical complementaries corresponding to the six primaries: from a primary, by way of neutral gray, to its complementary:

from red-violet to green; from orange to cyan; from yellow-green to blue-violet; from green to magenta; from blue-green to red-orange; from blue to yellow.

To Cut Out

Experiment for Yourself

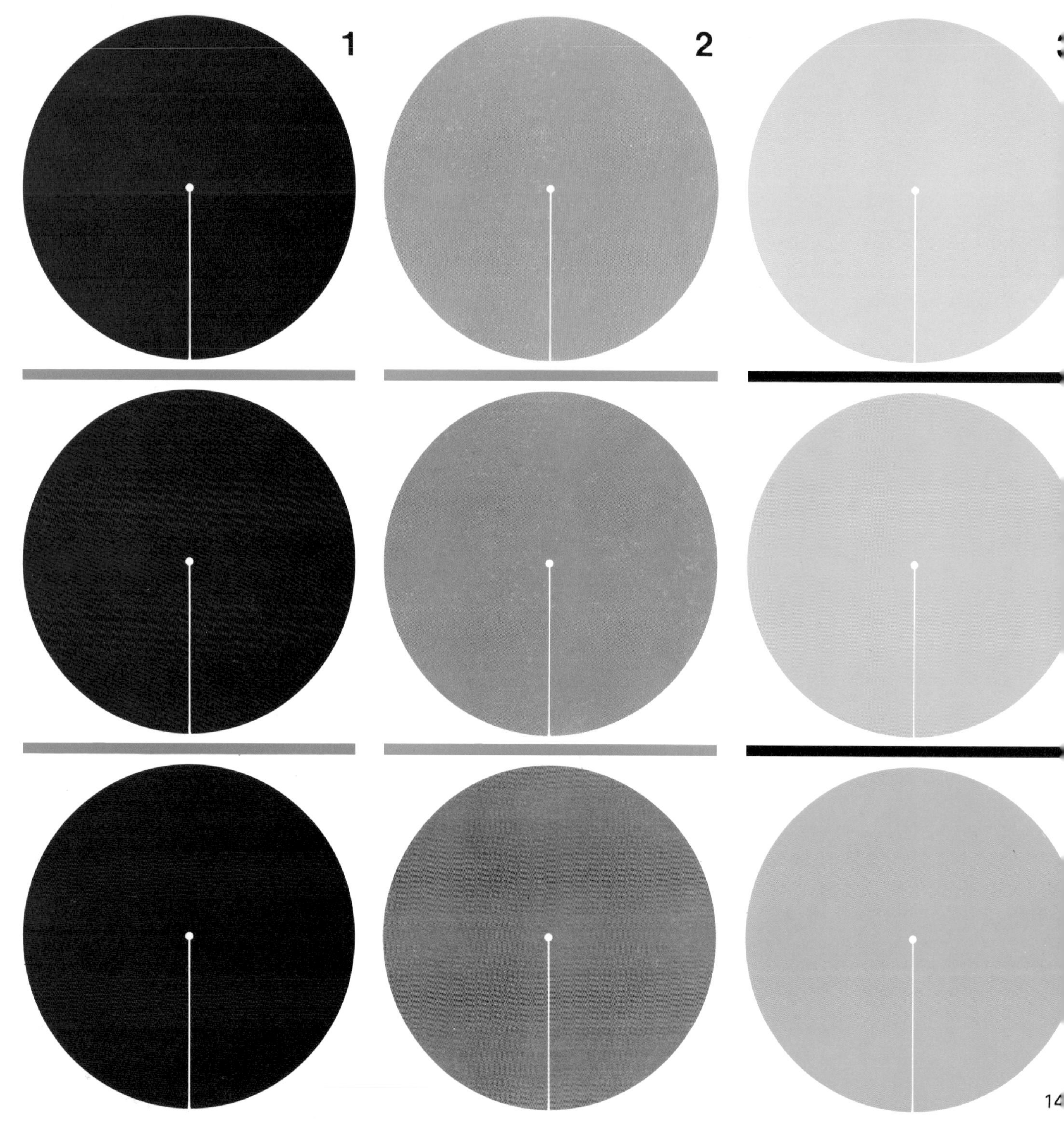
1
2
3

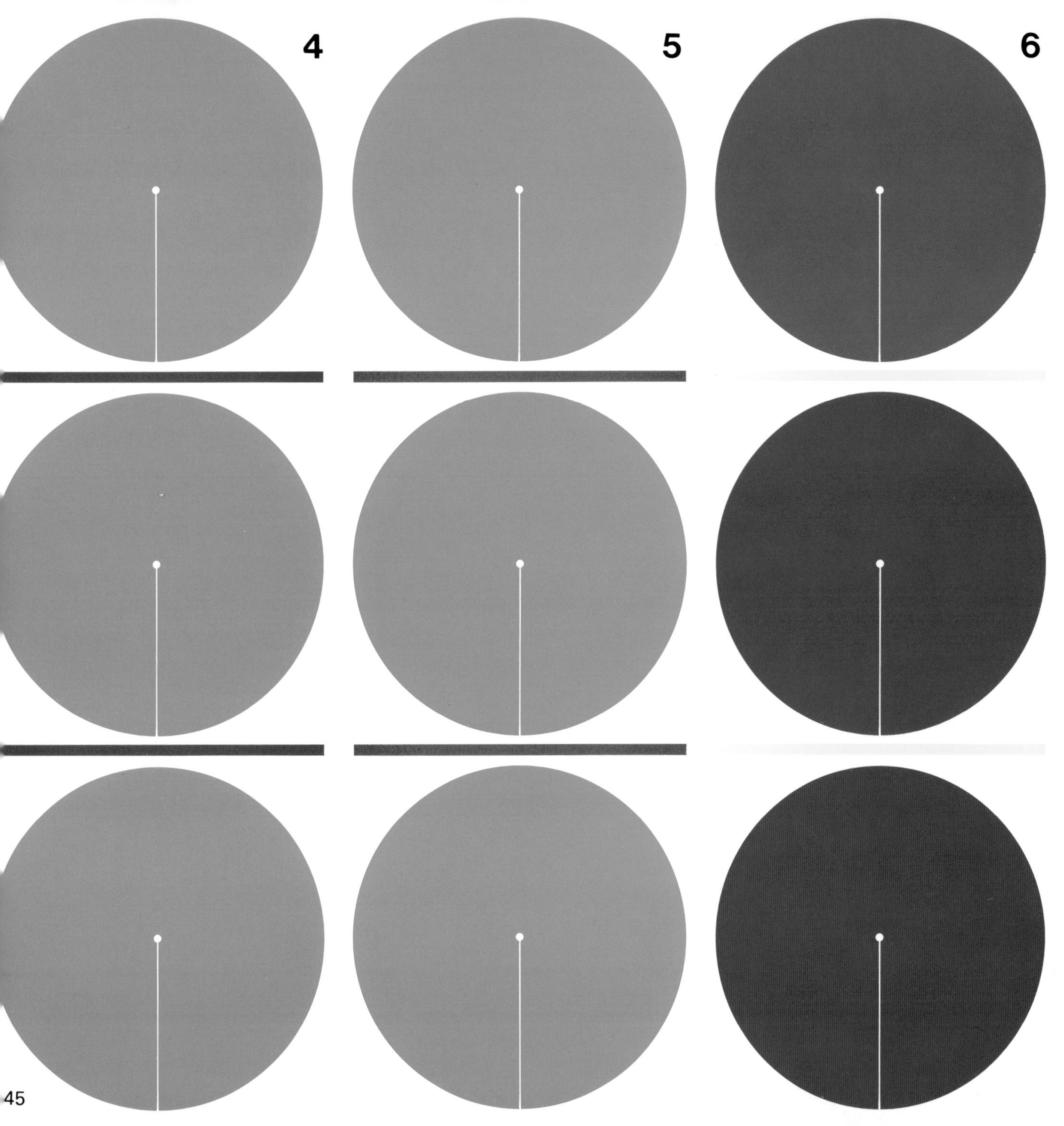
4
5
6

To Cut Out

Experiment for Yourself

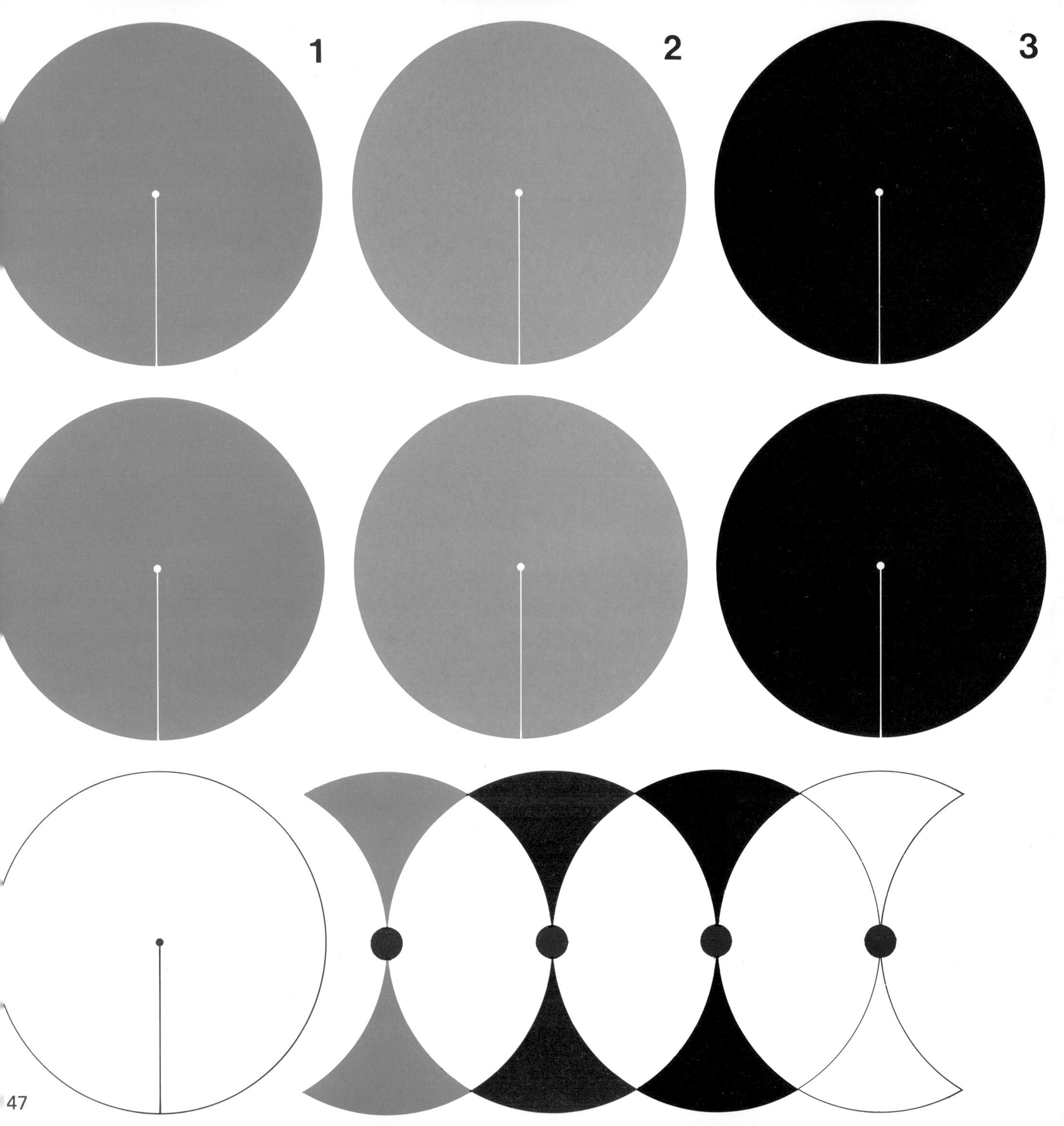
1
2
3

To Cut Out

Experiment for Yourself

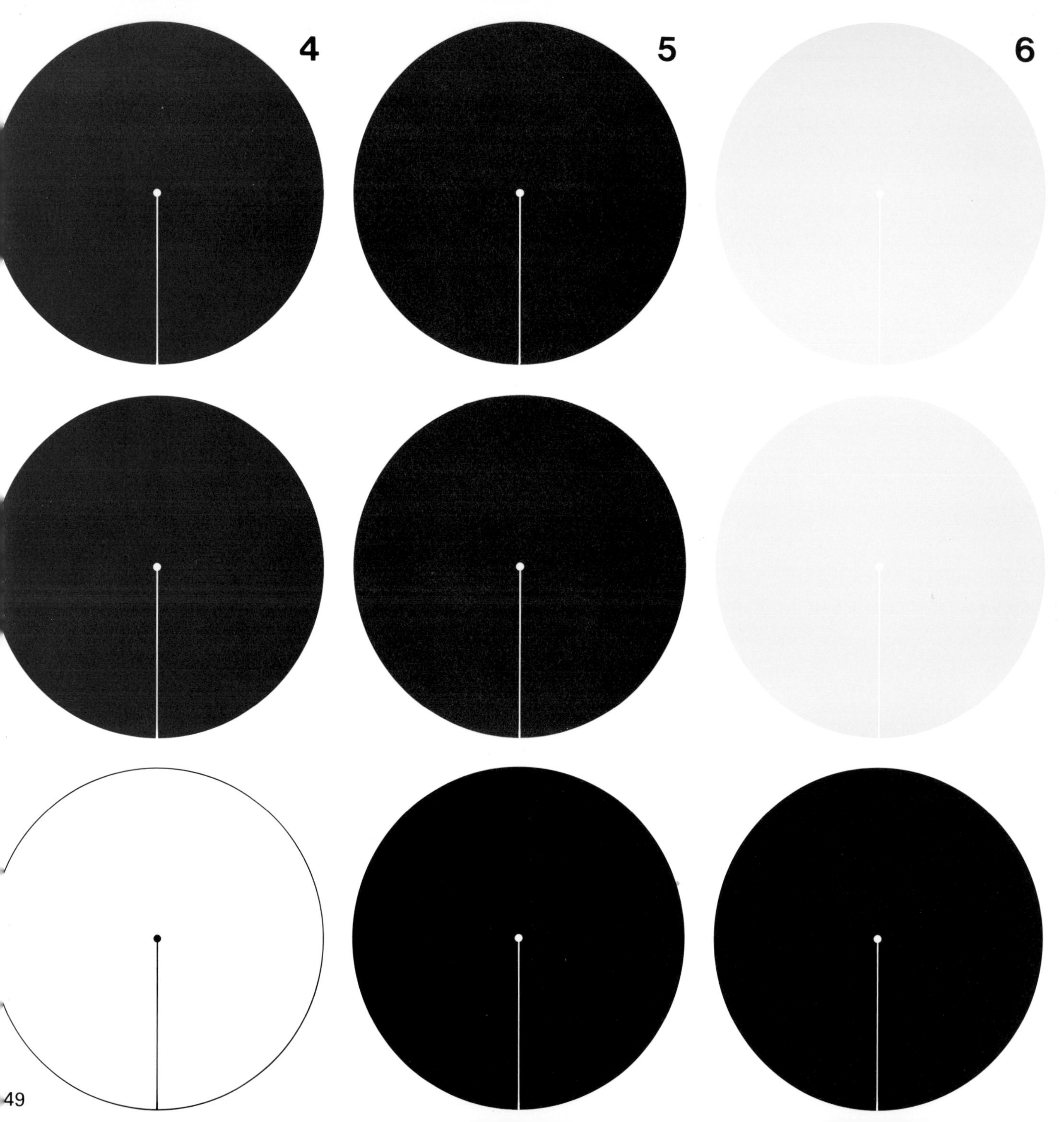
4
5
6

To Cut Out

Experiment for Yourself

Conclusion

The simultaneous phenomenon reveals our powers of differentiation, by increasing the contrasts between form and background, by relating an object, a surface, a color to its environment.

Optical mixing takes a heterogeneous structure and transforms it into a new uniform combination which is more easily discernible as an entity in a wider spatio-temporal situation.

Though these phenomena are opposite in their effect — simultaneity exaggerates the contrasts in the confrontation of two hues or two brightnesses, and optical mixing levels out the differences — they express a unity on the level of relationships of energy.

The coefficient of difference in the simultaneous effects varies between 30% and 50%. The gap becomes greater as the energetic contrasts juxtaposed are increased — culminating, in the blue/yellow and white/black pairs, in 50%.

For the optical average we have observed a parallel difference in the respective quantities which increases the more the brightness and dynamism of the colors in question are in opposition, varying from 24% to 50%, with the greatest reduction in difference occurring also with the blue/yellow and white/black pairs.

This reality takes us in the direction of the fundamental discovery of contemporary physics: no element in the universe exists by itself. It is in constant interaction with its close and distant neighbors. It takes on a meaning solely in relation to the location in space and time in which it is perceived by a consciousness.

References

1. Frans Gerritsen
   Présence de la couleur
   Dessain et Tolra, 1975

2. Edouard Fer
   Solfège de la couleur
   Dunod, 1970

3. H. Zollinger
   A correlation between Linguistics of Colour-naming and Colour Perception
   Adam Hilger, London 1973

4. Harald Küppers
   a. *La couleur*
   Dessain et Tolra 1975
   b. *Die Logik der Farbe*
   Callwey, 1976

5. R. L. Gregory
   a. *Eye and Brain*
   McGraw-Hill, 1966
   b. *The Intelligent Eye*
   McGraw-Hill, 1970

6. *Taschenlexikon der Farben*
   Kornerup und Wanscher
   Musterschmidt-Verlag, Göttingen, 1963

7. M. E. Chevreul
   Principles of Harmony and Contrast of Colors.
   Van Nostrand Reinhold, 1981 (first published 1839)

8. Ellen Marx
   Démonstration de la synthèse soustractive: Les contrastes de la couleur
   Dessain et Tolra, 1973

9. P. L. Walraven
   "Theoretical Models of the Colour Vision Network"
   in *Colour 73*
   Adam Hilger, London

   J. J. Vos
   The Zone-fluctuation Model of Color Vision
   AIC COLOR 1981, Berlin

   See also:

   Mac Nicol
   "Color Discrimination Processes in the Retina"
   Second Congress of International Color Association, New York University

10. Yves Le Grand and S. G. El Hage.
    Physiological Optics
    Springer-Verlag, 1980

11. Josef Albers
    Interaction of Color
    Yale University Press, 1975

12. Goethe, Johann V.
    Theory of Colours
    Du Mont, 1974

13. R. W. G. Hunt
    "Problems in colour reproduction"
    in *Colour 73,* page 125
    Adam Hilger, London

This pocket contains six transparent sheets (yellow, cyan, magenta) and instructions for using them, and a screen for successive projection.